NUREG-0693

I0482740

Analysis of Ultimate Heat Sink Cooling Ponds

R. Codell, W. K. Nuttle

**Office of
Nuclear Reactor Regulation**

**U.S. Nuclear Regulatory
Commission**

Analysis of Ultimate Heat Sink Cooling Ponds

Manuscript Completed: July 1980
Date Published: November 1980

R. Codell, W. K. Nuttle

Division of Engineering
Office of Nuclear Reactor Regulation
U.S. Nuclear Regulatory Commission
Washington, D.C. 20555

ABSTRACT

A method to analyze the performance of ultimate heat sink cooling ponds is presented. A simple mathematical model of a cooling pond is used to scan weather data to determine the period of the record for which the most adverse pond temperature or rate of evaporation would occur. Once the most adverse conditions have been determined, the peak pond temperature can be calculated. Several simple mathematical models of ponds are described; these could be used to determine peak pond temperature, using the identified meteorological record. Evaporative water loss may be found directly from the scanning by a simple and conservative heat-and-material balance.

Methodology by which short periods of onsite data can be compared with longer offsite records is developed, so that the adequacy of the offsite data for pond performance computations can be established.

CONTENTS

CONTENTS (Continued)

CONTENTS (Continued)

LIST OF FIGURES

LIST OF TABLES

SYMBOLS

A	pond surface area, ft^2 or acres
A_0	one-half the daily insolation, Btu/ft^2
A_n	surface area of nth segment of the plug-flow model, ft^2
C	cloud cover in tenths of the total sky obscured
C_1	Bowen's ratio, ~ 0.26 mm $Hg/°F$
C_p	heat capacity of water, $Btu/lb/°F$
E	equilibrium temperature, $°F$
E_1, E_2	estimation of equilibrium temperatures using data from offsite and onsite records, respectively, $°F$
$E(\bar{x})$	estimation of equilibrium temperature using monthly average meteorologic data, $°F$
e_a	saturation pressure of air above pond surface, mm Hg
e_s	saturation pressure of air at surface temperature T_S, mm Hg
g	skew coefficient
H	heat content, Btu
H_N	heat transfer from segment N, Btu/day
H_{vap}	heat of vaporization of water, Btu/lb
$\dot{H}$	net heat flux, $Btu/(ft^2 \; day)$
$\dot{H}_{AN}$	net atmospheric longwave radiation, $Btu/(ft^2 \; day)$
$\dot{H}_{BR}$	back radiation from pond surface, $Btu/(ft^2 \; day)$
$\dot{H}_C$	conductive and convective heat loss, $Btu/(ft^2 \; day)$
$\dot{H}_E$	evaporative heat loss, $Btu/(ft^2 \; day)$
$\dot{H}_n$	heat transfer from segment n, Btu/day
$\dot{H}_{RJ}$	net plant heat rejection, $Btu/(ft^2 \; day)$
$\dot{H}_S$	gross solar radiation
$\dot{H}_{SN}$	net solar radiation, $Btu/(ft^2 \; day)$
K	equilibrium heat transfer coefficient, $Btu/(ft^2 \; day°F)$
k	error band scale factor
M	sample mean
P	probability
P_∞	probability of occurrence for an event from an infinite population
P_N	probability of occurrence for an event from a finite population
p	atmospheric pressure, mm Hg
q	heat flow inside a pond, Btu/hr

ANALYSIS OF ULTIMATE HEAT SINK COOLING PONDS

1. INTRODUCTION

The ultimate heat sink (UHS) is defined as the complex of sources of service or house water supply necessary to safely operate, shut down, and cool down a nuclear power plant. Cooling ponds, spray ponds, and mechanical draft cooling towers are some examples of the types of ultimate heat sinks in use today.

The U.S. Nuclear Regulatory Commission (NRC) has set forth in Regulatory Guide 1.27 (Ref. 1) the following positions on the design of ultimate heat sinks:

(1) The ultimate heat sink must be able to dissipate the heat of a design-basis accident (e.g., loss-of-coolant accident) of one unit plus the heat of a safe shutdown and cooldown of all other units it serves.

(2) The heat sink must provide a 30-day supply of cooling water at or below the design-basis temperature for all safety-related equipment.

(3) The system must be shown to be capable of performing under the meteorologic conditions leading to the worst cooling performance and under the conditions leading to the highest water loss.

This report identifies methods that may be used to select the most severe combinations of controlling meteorologic parameters for surface cooling pond heat transfer and evaporative water loss. The procedure scans a long weather record, which is usually available from the National Weather Service for a nearby station, and it predicts the period for which either pond temperature or water loss would be maximized for a hydraulically simple cooling pond. The principle of linear superposition is assumed, which allows the peak ambient pond temperature to be superimposed on the peak "excess" temperature due to plant heat rejection. This procedure determines the timing within the weather record of the peak ambient pond temperature. The true peak can then be determined in a subsequent, more rigorous calculation.

Maximum evaporative water loss is determined by picking the 30-day continuous period of the record which has the highest evaporation losses and assuming that all heat rejected by the plant results in the evaporation of pond water.

To be effective the data scanning procedure requires a data record on the order of tens of years in length. Since these data will usually come from somewhere other than the site itself (such as an airport), methods to compare these data with the limited onsite data are developed so that the adequacy, or at least the conservatism, of the offsite data can be established. Conservative correction factors to be added to the final results are suggested.

These models and methods are provided as useful tools for UHS analyses of cooling ponds. They are intended as guidelines only. Use of these methods does not automatically assure NRC approval, nor are they required procedures for nuclear power plant licensing. Furthermore, by publishing this guidance NRC does not wish to discourage independent assessments of UHS performance or the furtherance of the state of the art.

2. HEAT AND MASS TRANSFER RELATIONSHIPS IN PONDS

The relationship used in this report for the transfer of heat and water vapor from the pond surface is developed along the lines of the "equilibrium temperature" procedure of Brady et al. (Ref. 2) and Edinger et al. (Ref. 3). The main reasons for the choice of this procedure are:

- It is inherently simple.

- It can be shown to be conservative.

- It makes possible visualization of the concept of "excess temperature."

This last point serves as a basis for the separation of the pond temperature responses as a result of environmental forces from those which result from plant driving forces; this separation further facilitates the scanning of weather data as described below.

Other heat transfer relationships may be more accurate than the one used here; however, the selection of the period of meteorological record giving the most adverse pond temperature or evaporation should be fairly insensitive to the heat transfer relationship or pond model. Therefore, it is acceptable to use the proposed heat transfer and pond hydraulic model to scan the weather record, and then to use that record with a more sophisticated heat transfer and pond model for final determination of the maximum pond temperature and water losses.

2.1 Equilibrium Temperature Heat Transfer Model

The temperature the pond would reach at steady state without external heat inputs and under constant environmental conditions is known as the equilibrium temperature E. The equilibrium temperature is the temperature at which the heat removal from the pond balances the heat addition. This relation is graphically illustrated in Figure 2.1. Equilibrium temperature, therefore, is a rigorously definable property, dependent on the meteorological conditions at an instant in time. The equilibrium heat transfer coefficient K is also illustrated in Figure 2.1 and is defined as the slope of the heat removal curve at pond temperature $T_S = E$ for a unit surface area:

$$K = \left. \frac{\partial \dot{H}}{\partial T} \right)_E \tag{2-1}$$

3

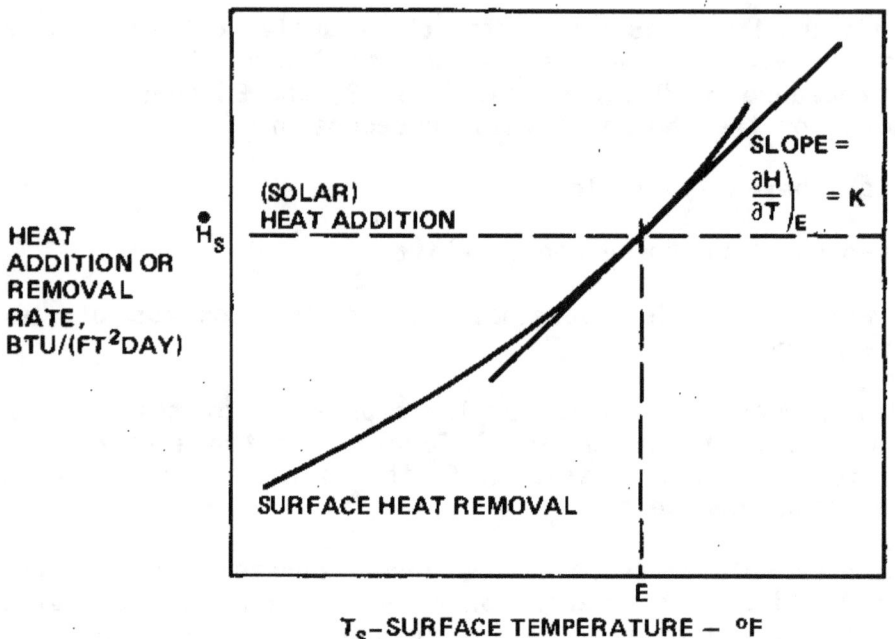

Figure 2.1 Definition of Equilibrium Coefficients.

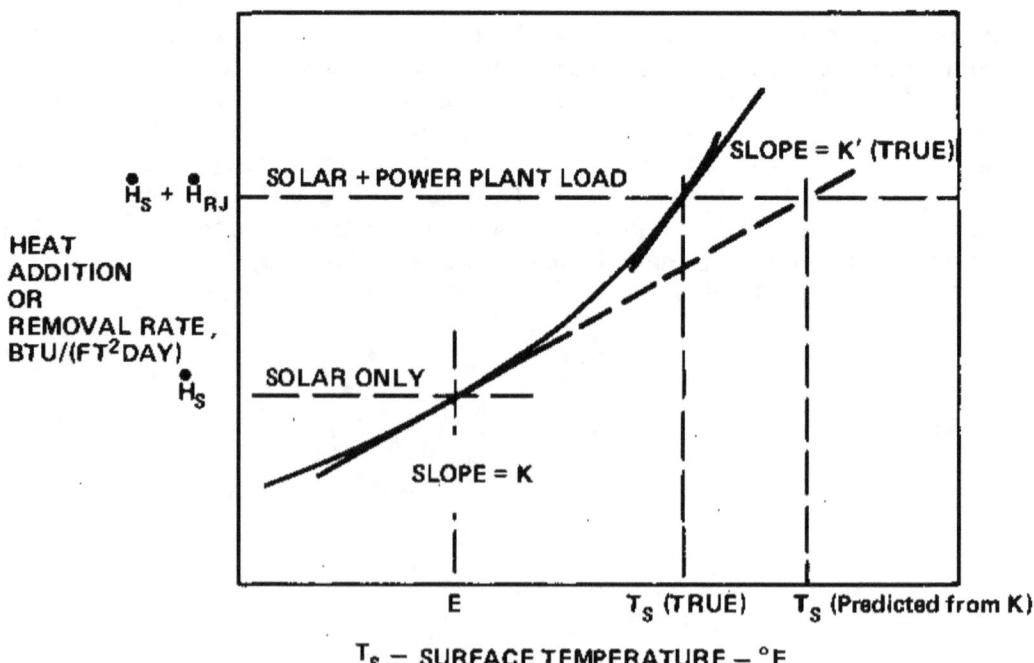

Figure 2.2 Pond Temperature Computations (Steady State).

4

In this case, the heat transfer $\dot{H}$ can be described as

$$\int_0^{\dot{H}} d\dot{H} = \int_E^{T_S} K \, dT, \qquad\qquad (2\text{-}2)$$

providing that K is reasonably constant in the interval T_S to E. This is, of course, approximately true only if T_S is very close to E. In a thermally heavily loaded pond, this assumption is not correct. The heat removal curve has an increasing slope at higher pond temperatures; therefore, the heat transfer may be underestimated for high heat loadings, and the predicted pond temperature may be too high. This potential error is shown graphically in Figure 2.2. The external heat load on the pond must, therefore, be factored into the determination of pond temperature.

2.2 Development of the Basis for Surface Heat and Mass Transfer From a Pond

Mechanisms of surface heat and mass transfer have been extensively studied in connection with large, lightly loaded bodies of water, such as lakes and reservoirs. Much less work exists on small, heavily loaded ponds. Application of results from large water bodies must be applied to small, heavily loaded ponds cautiously and conservatively until further experimental evidence of pond performance can be gathered. The NRC is sponsoring experiments on such ponds with Battelle, Pacific Northwest Laboratories.

A relationship for the rate of net heat flow into the pond can be developed through consideration of each heat source and heat loss. It is assumed that all heat exchange with an isolated body of water takes place through its surface. The rate of heat exchange $\dot{H}$ is

$$\dot{H} = \dot{H}_{SN} + \dot{H}_{AN} - \dot{H}_{BR} - \dot{H}_E - \dot{H}_C + \dot{H}_{RJ} \qquad \text{Btu/(ft}^2 \text{ day)} \qquad (2\text{-}3)$$

in which:

$\dot{H}$ = net rate of heat flow into the pond

$\dot{H}_{SN}$ = net rate of shortwave solar radiation entering the pond, measured directly

$\dot{H}_{AN}$ = net rate of longwave atmospheric radiation entering the pond, measured directly

$\dot{H}_{BR}$ = net rate of back radiation leaving the pond surface

$\dot{H}_E$ = net rate of heat loss due to evaporation

$\dot{H}_C$ = net rate of heat flow from the pond due to conduction and convection

$\dot{H}_{RJ}$ = net rate of heat addition by the plant

This relationship is illustrated graphically in Figure 2.3.

5

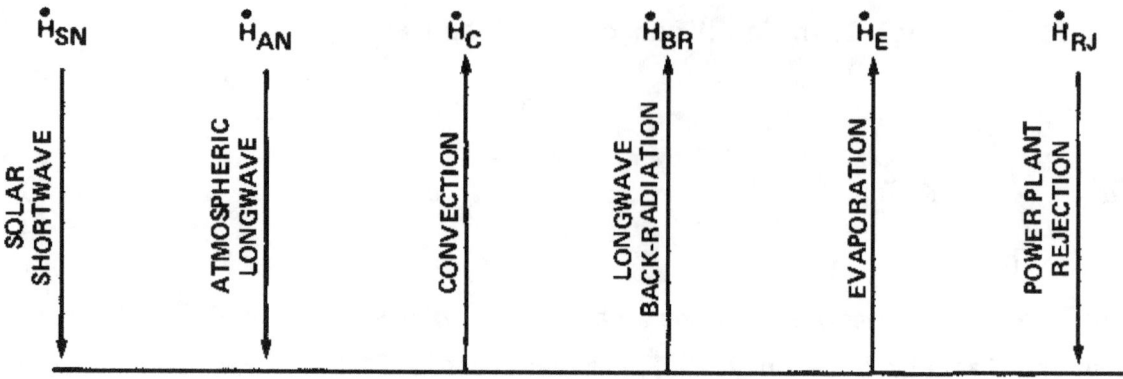

Figure 2.3 Heat Loads on a Pond.

Of the heat flows into the pond as a result of radiation $\dot{H}_{SN}$ and $\dot{H}_{AN}$, only the net atmospheric radiation can be estimated from meteorologic parameters. The net atmospheric radiation term can be approximated using air temperature T_A and cloud cover C (in tenths). Ryan and Harleman (Ref. 4) developed the following formula for $\dot{H}_{AN}$:

$$\dot{H}_{AN} = 1.2 \cdot 10^{-13} (T_A + 460)^6 (1 + 0.17C^2) \qquad \text{Btu/(ft}^2 \text{ day)} \quad (2\text{-}4)$$

Three components of the heat exchange equation, $\dot{H}_{BR}$, $\dot{H}_E$, and $\dot{H}_C$, are functions of the pond surface temperature. The back radiation term may be expressed using the relation for radiation from a black body (Ref. 2):

$$\dot{H}_{BR} = 4.026 \times 10^{-8}(460 + T_S)^4 \qquad \text{Btu/(ft}^2 \text{ day)} \qquad (2\text{-}5)$$

where T_S is the surface temperature of the pond. Using the linear terms of the Taylor series expansion of this relation gives

$$\dot{H}_{BR} = 1801 + 15.7(T_S) \qquad \text{Btu/(ft}^2 \text{ day)} \qquad (2\text{-}6)$$

The evaporative heat flow can be estimated by

$$\dot{H}_E = (e_s - e_a)f(U) \qquad \text{Btu/(ft}^2 \text{ day)} \qquad (2\text{-}7)$$

in which e_s is the saturation vapor pressure at the temperature of the water surface (mm Hg) and e_a is the saturation vapor pressure of the air above the pond. The second term on the right is an empirical function of windspeed in miles per hour, U. The wind function proposed by Brady (Ref. 2) is used:

$$f(U) = 70 + 0.7U^2 \qquad \text{Btu/(ft}^2 \text{ day)/mm Hg} \qquad (2\text{-}8)$$

6

where U is measured at the 18-foot level.

The quantity $(e_s - e_a)$ can be replaced by a relationship using the slope of the vapor pressure versus temperature curve for some temperature between the surface temperature of the pond and the dew point temperature T_D,

$$(e_s - e_a) = \beta (T_S - T_D) \qquad \text{mm Hg} \qquad (2\text{-}9)$$

The following polynomial for β can be used in the temperature range normally encountered (Ref. 2):

$$\beta = 0.255 - 0.0085T^* + 0.00204(T^*)^2 \qquad \text{mm Hg/}°\text{F} \qquad (2\text{-}10)$$

where

$$T^* = \frac{T_S + T_D}{2}$$

Making the appropriate substitutions,

$$\dot{H}_E = \beta (T_S - T_D)f(U) \qquad \text{Btu/(ft}^2 \text{ day)} \qquad (2\text{-}11)$$

The conduction and convection heat flow can be approximated by

$$\dot{H}_C = C_1 (T_S - T_A)f(U) \qquad \text{Btu/(ft}^2 \text{ day)} \qquad (2\text{-}12)$$

where

T_A = air temperature

C_1 = Bowen's coefficient, 0.26 mm Hg/°F

Making the appropriate substitutions in Eq. (2-3), and neglecting the plant heat load for now, leads to

$$\dot{H} = H_{SN} + 1.2 \times 10^{-13}(T_A + 460)^6(1 + 0.17C^2) - (1801 + 15.7T_S)$$

$$- \beta(T_S - T_D)f(U) - 0.26(T_S - T_A)f(U) \qquad \text{Btu/(ft}^2 \text{ day)} \quad (2\text{-}13)$$

7

Equation (2-13) can be put into the equilibrium temperature form,

$$\dot{H} = K(E - T_S) \qquad \text{Btu/(ft}^2\text{ day)} \qquad (2\text{-}14)$$

Equation (2-13) is solved for K by letting $T_S = E$ and $\dot{H} = 0$:

$$0 = \dot{H}_{SN} + 1.2 \times 10^{-13}(T_A + 460)^6 (1 + 0.17C^2) - (1801 + 15.7E)$$
$$- \beta (E - T_D)f(U) - 0.26(E - T_A)f(U) \qquad (2\text{-}15)$$

Subtracting Eq. (2-15) from Eq. (2-13) gives

$$\dot{H} = 15.7(E - T_S) + (\beta + 0.26)(E - T_S)f(U) \qquad \text{Btu/(ft}^2\text{ day)} \qquad (2\text{-}16)$$

Comparison with Eq. (2-14) leads to a relation for K:

$$K = 15.7 + (\beta + 0.26)f(U) \qquad \text{Btu/(ft}^2\text{ day}^\circ\text{F)} \qquad (2\text{-}17)$$

The pond is likely to have its lowest cooling capacity during the summer months, since ambient temperatures will be higher. It can be shown, using Eqs. (2-4) and (2-5), that the components of atmospheric radiation and back radiation from the pond surface nearly balance in the warmer months. The error of neglecting both terms under these conditions is small. The elimination of the atmospheric- and back-radiation terms from Eq. (2-13) allows for the explicit solution for the equilibrium temperature. Equation (2-13) becomes

$$\dot{H} = \dot{H}_{SN} - \beta(T_S - T_D)f(U) - 0.26(T_S - T_A)f(U) \qquad \text{Btu/(ft}^2\text{ day)} \quad (2\text{-}18)$$

Substituting $E = T_S$, $\dot{H} = 0$, and solving for E gives

$$E = \frac{\dot{H}_{SN}}{(\beta + 0.26)f(U)} + \frac{(\beta T_D + 0.26T_A)}{(\beta + 0.26)} \qquad ^\circ\text{F} \qquad (2\text{-}19)$$

Equating Eq. (2-14) with Eq. (2-18),

$$K(E - T_S) = \dot{H}_{SN} - \beta(T_S - T_D)f(U) - 0.26(T_S - T_A)f(U) \qquad (2\text{-}20)$$

Substituting from Eq. (2-19) for E and solving for K gives

$$K = (\beta + 0.26)f(U) \qquad \text{Btu/(ft}^2 \text{ day } ^\circ\text{F)} \qquad (2\text{-}21)$$

This allows Eq. (2-19) to be put in its final form,

$$E = \frac{\dot{H}_{SN}}{K} + \frac{(\beta T_D + 0.26T_A)}{(\beta + 0.26)} \qquad ^\circ\text{F} \qquad (2\text{-}22)$$

Equation (2-23) is the alternate formulation of Eq. (2-17) which follows from the approximation $H_{AN} \sim H_{BR}$. So, the surface heat transfer equation as it is used in the model has the form

$$\dot{H} = K(E - T_S) \qquad \text{Btu/(ft}^2 \text{ day)} \qquad (2\text{-}14)$$

$$K = 15.7 + (\beta + 0.26)f(U) \qquad \text{Btu/(ft}^2 \text{ day)} \qquad (2\text{-}17)$$

$$\beta = 0.255 - 0.0085T^* + 0.00204(T^*)^2 \qquad \text{mm Hg/}^\circ\text{F} \qquad (2\text{-}10)$$

$$T^* = \frac{T_S + T_D}{2} \qquad ^\circ\text{F}$$

$$f(U) = 70 + 0.7U^2 \qquad \text{Btu/(ft}^2 \text{ day)/mm Hg} \qquad (2\text{-}8)$$

$$E = \frac{\dot{H}_{SN}}{K} + \frac{(\beta T_D + 0.26T_A)}{(\beta + 0.26)} \qquad ^\circ\text{F} \qquad (2\text{-}22)$$

in which

T_S = pond surface temperature, °F

T_A = air temperature, °F

T_D = dew point temperature, °F

U = windspeed, mph, measured at the 18-foot level

$\dot{H}_{SN}$ = net short wave solar radiation received by the pond, Btu/(ft² day)

Evaporation is calculated directly from the evaporative heat flux:

$$W_e = \frac{\beta(T_S - T_D)f(U)}{\rho\, H_{vap}}$$ (2-23)

where

W_e = evaporative flux per unit area of surface   ft³/hr/ft²

H_{vap} = heat of vaporization of water   Btu/lb

2.3 Conservatism of Equilibrium Temperature Formulation

The formulation of the heat transfer formulae used has a number of builtin conservatisms, which tend to overestimate pond temperature. One of the larger conservatisms is the choice of a wind dependence $f(U)$. The Brady wind function employed seems to underestimate the evaporative flux, even when compared to Brady's own data (Ref. 4).

Brady's wind function is derived empirically from large lake data. A more accurate, but less conservative formula was derived by Ryan (Ref. 4) on firmer physical grounds:

$$f(U_2) = [22.4\ (\Delta\theta_v)^{1/3} + 14U_2)$$ (2-24)

$$\Delta\theta_v = \frac{T_S + 460}{1 - \dfrac{0.378e_s}{p}} - \frac{T_A + 460}{1 - \dfrac{0.378e_a}{p}}$$

where U_2 is expressed in mph measured 2 m above water surface, and

$\Delta\theta_v$ = virtual temperature, °F

and where

p = atmospheric pressure, mm Hg

10

This formula accounts for an expected increase in natural convection with increasing pond temperature, whereas Brady's wind function is not temperature dependent.

A simple example illustrates the different heat transfer formulations and how they can drastically affect the temperature calculations. The parameters refer to a one-square-foot section of pond surface:

Solar input = 2100 Btu/(ft^2 day)
Dew point temperature = 70°F
Ambient air temperature = 90°F
Windspeed = 2 mph
Power plant load = 0 to 11,000 Btu/(ft^2 day)

Four heat transfer formulas are used to calculate the steady-state pond temperature in response to these meteorological parameters:

(1) The equilibrium temperature and heat transfer coefficients based on unloaded pond conditions (not a function of pond temperature).

(2) The equilibrium temperature and heat transfer coefficients based on pond temperature (method used in present models).

(3) Rigorous formula—each of heat transfer terms in Eq. (2-3) is explicitly calculated with Brady wind function used.

(4) Rigorous formula—same as case 3 but with Ryan wind function used.

The results of this calculation are presented in Figure 2.4. Although all of the four formulas are in good agreement at light pond loadings, they deviate substantially at high loadings. The conservatism of the Brady wind function over the Ryan wind function is evident, as is the conservatism of the approximation formula over the rigorous formula.

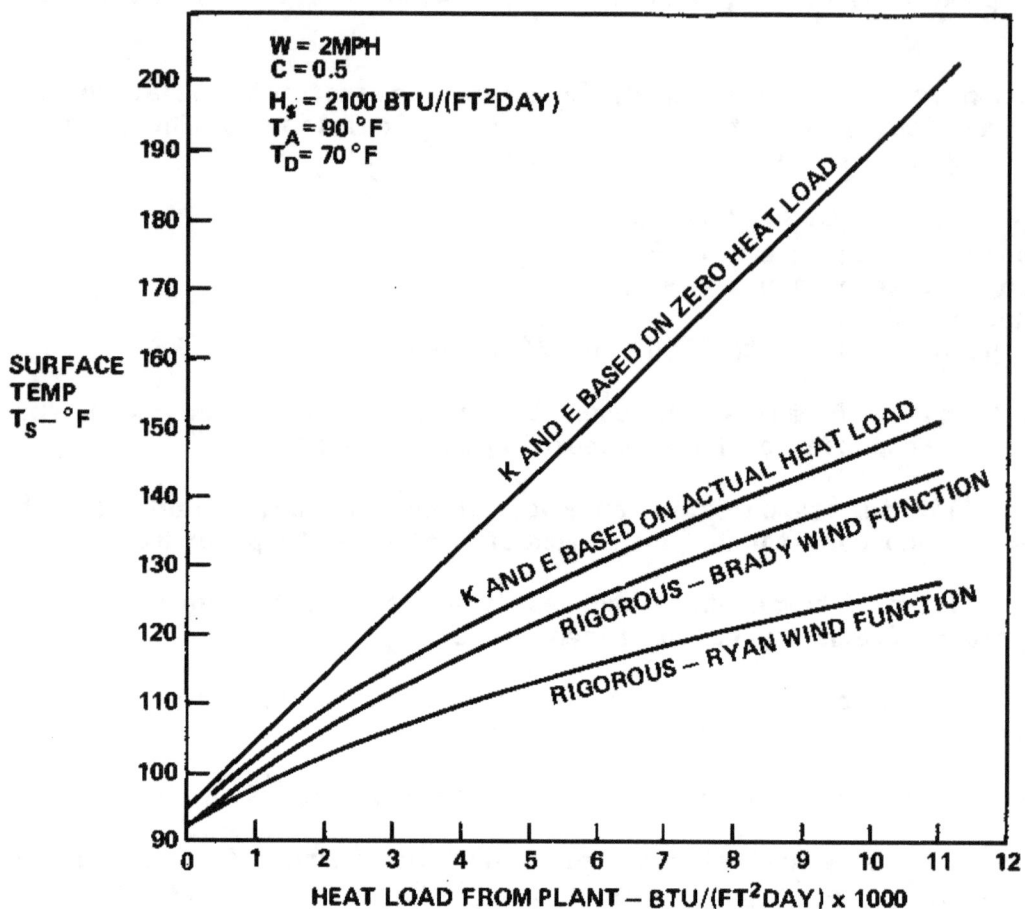

Figure 2.4 Steady-State Surface Temperatures.

12

3. POND MODELS

The hydrodynamics of small cooling ponds typically used for ultimate heat sinks can be extremely complicated. Because of its lower density, heated water may stratify on the surface of the pond. In many cases, this stratification may be used to good advantage in order to segregate the heated upper water layer from the cooler underlying water. This may be accomplished by designing and locating the intake from and discharges to the pond so as to minimize mixing. Mixing of the waters in the pond because of improper design, or mixing induced by high winds will tend to lower the efficiency of the pond by "short-circuiting" the hot and cool layers.

Mathematical models capable of accurately simulating thermal hydraulics of ponds are becoming increasingly available (Refs. 5 and 6). The methods described here, however, involve only very simple, idealized pond models. Complicated mathematical models are only as good as their data input and the ability of the modeler to describe the pond and its environment. It has not been found necessary to employ sophisticated pond models to the meteorological screening procedure.

Three of the simple models will be described.

(1) The mixed-tank model assumes total mixing of all heated effluent through-out the volume of the pond.

(2) The stratified-flow model assumes complete density stratification with the heated effluent entering the surface layer and the cooled water being withdrawn from the bottom layer.

(3) The plug-flow model assumes that the thermal effluent is discharged to the pond and travels as a "plug" through the entire volume of the pond, all the while transferring heat to the atmosphere.

Only the mixed-tank model is used in the data scanning procedure.

A well designed cooling pond will have hydraulic properties approaching those of the stratified- or plug-flow models, which are most efficient at dissipating heat (heated and cooled water do not mix in these designs). The mixed-tank model represents an inefficient design. Heated water entering the pond will be completely and instantly mixed with the total pond inventory. Therefore, part of the heated water will be recirculated before it has had the opportunity to be cooled; but part of the water will stay in the pond for a long period of time.

It is not necessarily true, however, that any of the three models would represent the prototype. In some cases, there could be conditions in the prototype pond that would lead to less efficient operation than predicted by any of the three models described here. For example, "side arms" of irregularly shaped reservoirs used for power plant cooling may be less efficient at rejecting heat than is the main body of the reservoir because circulation in these regions can be poor. Also, there exists the possibility that stratification may cause short-circuiting between the intake and discharge which would tend to isolate

13

the cold water of the lower layer from the intake and thereby effectively reduce
the thermal inertia of the pond, that is, its capacity to absorb initially high
heat loads. For example, see the analysis in Jirka (Ref. 7).

These pond models do not explicitly simulate the complicated hydrodynamic features
of ponds. If the possibility of factors that would reduce efficiency exist,
arbitrary reductions of surface area and pond volume can be made to assure conservatism. Furthermore, the relative simplicity of the models allows their
incorporation into a pond temperature computer program "UHS3," to be described
later, in which all three pond configurations are considered simultaneously.

3.1 Mixed-Tank Model

The mixed-tank model depicted in Figure 3.1 presumes that the heated effluent
is instantaneously and uniformly mixed throughout the volume of the tank, and
that the water in the tank is uniform in temperature. Heat transfer with the
atmosphere occurs at the surface of the tank. This heat transfer is less
efficient in the case of a completely mixed pond than in a pond where the hot
and cool water do not mix. Atmospheric heat transfer is related to the pond
surface temperature, which is diminished in the former case and preserved in
the latter.

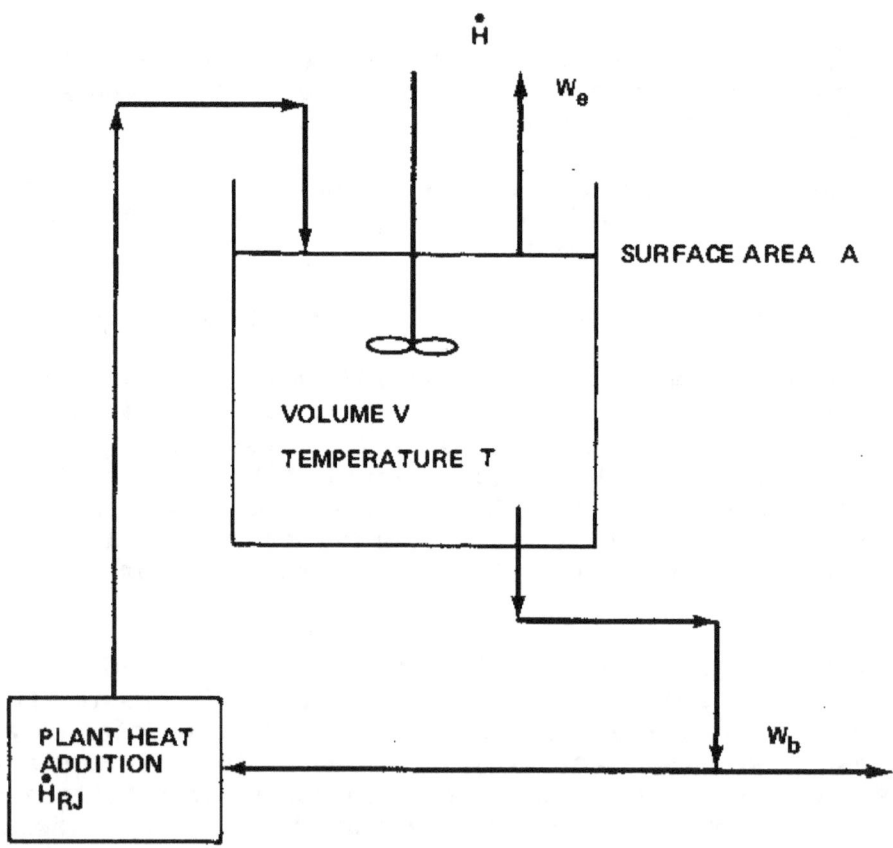

Figure 3.1 Mixed-Tank Model.

14

3.1.1 Heat Balance

A heat and mass balance can be formulated for the mixed-tank model. The terms of the heat balance are:

3.1.1.1 Heat Load Into Pond

$$\text{Heat in} = \dot{H}_{RJ} \qquad \text{Btu/hr} \qquad\qquad (3\text{-}1)$$

3.1.1.2 Heat Out From Surface

$$\dot{H} = \frac{AK\ (T\text{-}E)}{24} \qquad \text{Btu/hr} \qquad\qquad (3\text{-}2)$$

where

A = surface area of the pond, ft^2.
K = equilibrium heat transfer coefficient, $Btu/(ft^2\ day)/°F$
T = mixed-pond temperature, $°F$
E = equilibrium temperature, $°F$

3.1.1.3 Heat Out in Blowdown or Leakage Stream

With reference to the mixed-pond temperature T, heat loss from blowdown is by definition zero:

$$q_b = W_b C_p\ (T - T) = 0 \qquad\qquad (3\text{-}3)$$

where

W_b = flowrate of the blowdown or leakage stream

ρ = density of water, lb/ft^3

C_p = specific heat of water, $Btu/lb/°F$

Combining all heat inputs to and outputs from the pond, and using the relationship relating temperature to heat, the following is obtained:

$$\frac{dT}{dt} = \frac{\dot{H}_{RJ}}{C_p\ V} - \frac{AK}{24\ C_p\ V}\ (T - E) \qquad °F/hr \qquad\qquad (3\text{-}4)$$

where V is the pond volume.

3.1.2 Mass Balance

The mass balance on the pond includes evaporative loss from the surface and the blowdown or leakage. The terms of the mass balance are:

15

Blowdown on leakage flow = W   ft^3/hr

Evaporative loss from surface = W_e ft^3/hr/ft^2

$$W_e = \frac{\beta(T_S - T_D)\ f(U)\ A}{24\ \rho\ H_{vap}} \qquad (3-5)$$

where

β = slope of the vapor pressure-temperature curve, mm Hg/°F

T_D = dew point temperature, °F

H_{vap} = heat of vaporization of water, Btu/lb

Combining all terms of the mass balance yields the expression:

$$\frac{dV}{dt} = -W_b \quad \frac{-\ \beta(T_S - T_D)\ f*U)\ A}{24\ \rho\ C_p\ H_{vap}} \qquad \text{ft}^3 \qquad (3-6)$$

3.2 Stratified-Flow Model

In the stratified-flow model (Figure 3.2) the assumption is made that the heated effluent enters on the surface of the pond and cooled water is withdrawn from the bottom of the pond. Water does not mix vertically but simply moves from the surface layer to the bottom as a "plug." Heat transfer occurs only from the surface layer.

The pond is segmented into N horizontal slices of thickness ΔZ as shown in Figure 3.2. In the computer program subsequently described, N = 10. The terms of the energy balance of segment n are:

3.2.1 Heat Balance

3.2.1.1 Heat Entering From Above

$$q_{n-1} = W\ \rho\ C_p\ T_{n-1} \qquad \text{Btu/hr} \qquad (3-7)$$

where

W = flowrate through pond   ft^3/hr

3.2.1.2 Heat Leaving From Segment by Advection

$$q_n = W\ \rho\ C_p\ T_n \qquad \text{Btu/hr} \qquad (3-8)$$

3.2.1.3 Change in Heat Content During Time Δt

$$\Delta H = A\ \Delta Z\ \rho\ C_p\ \Delta T \qquad \text{Btu/hr} \qquad (3-9)$$

16

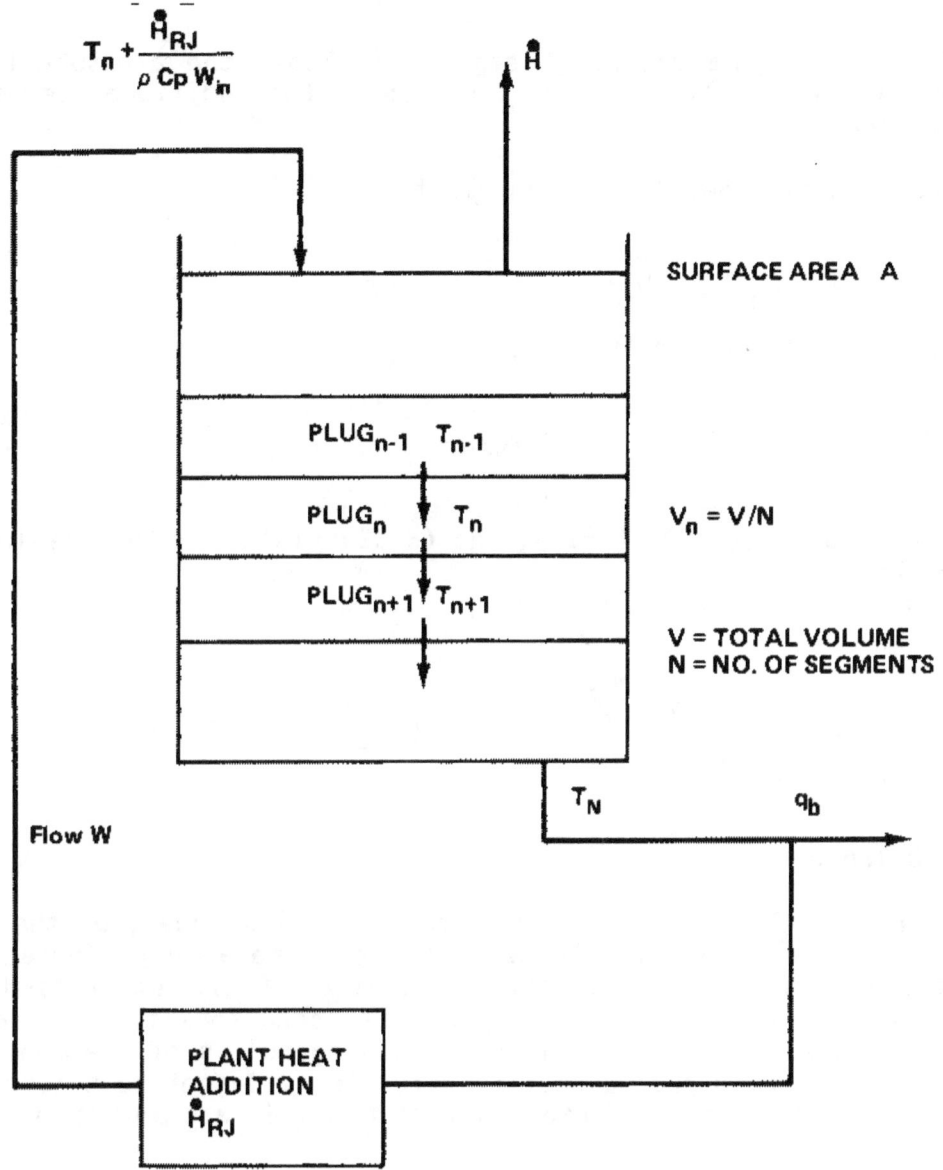

Figure 3.2 Stratified-Flow Model.

Combining all terms yields the following expression for segment n:

$$\frac{\Delta T_n}{\Delta t} = \frac{W}{A} \frac{(T_{n-1} - T_n)}{\Delta Z} \qquad °F/hr \qquad (3\text{-}10)$$

17

The heat balance of the uppermost segment includes the atmospheric heat transfer and the heat addition from the plant. The additional terms of this heat balance are:

3.2.1.4 Heat Entering Segment 1 From Plant

$$q_0 = (W \rho C_p T_N + \dot{H}_{RJ}) \qquad \text{Btu/hr} \qquad (3\text{-}11)$$

3.2.1.5 Heat Transferred to Atmosphere

$$\dot{H} = \frac{KA}{24}(T_1 - E) \qquad \text{Btu/hr} \qquad (3\text{-}12)$$

combining Eqs. (3-8), (3-9), (3-11), and (3-12) yields the expression for the first segment:

$$\frac{\Delta T_1}{\Delta t} = \frac{W}{A}\frac{(T_N - T_1)}{\Delta Z} + \frac{\dfrac{\dot{H}_{RJ}}{24} - KA(T_1 - E)}{\rho C_p A \Delta Z} \qquad °F/hr \qquad (3\text{-}13)$$

3.2.2 Mass Balance

No mass balance is formulated for the stratified-flow model, or the plug-flow model subsequently described. Instead, the mass balance performed on the mixed-tank model is used to correct the volume of the stratified- and plug-flow models. This approach is justified because the amount of heat leaving the pond surface is roughly the same regardless of the model chosen. Since 50% to 80% of atmospheric heat transfer is by latent heat (evaporation), the consumptive water use predicted by each model is assumed to be about the same.

3.3 Plug-Flow Model

The plug-flow model depicted in Figure 3.3 assumes that the heated effluent enters the pond and travels horizontally as a "plug" of water through the entire volume of the pond, exchanging heat to the atmosphere. Water in the plug does not mix horizontally, but the temperature in the plug is assumed to be uniform vertically.

The pond is segmented into N vertical slices of length ΔX as depicted in Figure 3.3. In the computer program subsequently described, N = 10. The terms of the energy budget on segment n are:

3.3.1 Heat Balance

3.3.1.1 Heat Entering by Advection From Previous Segment

$$q_{n-1} = W \rho C_p T_{n-1} \qquad \text{Btu/hr} \qquad (3\text{-}14)$$

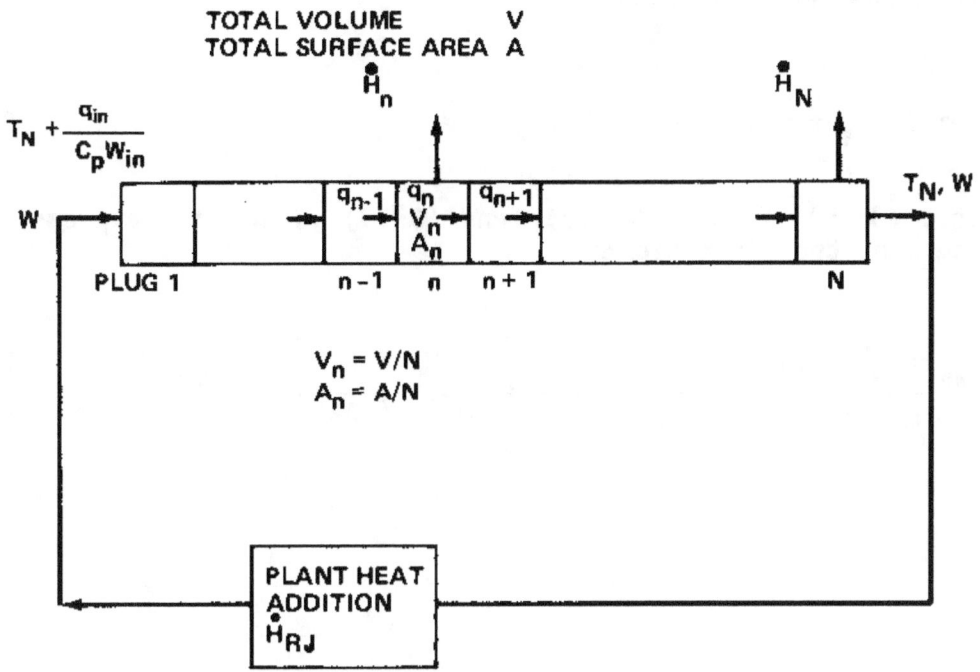

TOTAL VOLUME $\quad$ V
TOTAL SURFACE AREA $\quad$ A

$$V_n = V/N$$
$$A_n = A/N$$

Figure 3.3 Plug-Flow Model.

3.3.1.2 Heat Leaving Segment by Advection

$$q_n = W \, \rho \, C_p T_n \qquad \text{Btu/hr} \qquad (3\text{-}15)$$

3.3.1.3 Heat Transfer to Atmosphere From Segment

$$\dot{H} = \frac{KA}{24\,N} (T_n - E) \qquad \text{Btu/hr,} \qquad (3\text{-}16)$$

3.3.1.4 Change in Segment Heat Content During Time Δt

$$\Delta H = \frac{A\Delta X}{N} (\rho \, C_p \, \Delta T) \qquad \text{Btu/hr} \qquad (3\text{-}17)$$

combining all terms yields the expression for the temperature of segment n:

$$\frac{\Delta T_n}{\Delta t} = \frac{WN}{A} \, \frac{(T_{n-1} - T_n)}{\Delta X} - \frac{K \, (T_n - E)}{24 \, \rho \, C_p \, \Delta X} \qquad \degree\text{F/hr} \qquad (3\text{-}18)$$

The heat balance on the first segment includes the heat released from the plant and the recirculated heat from the pond. The additional term in this heat balance is:

19

3.3.1.5 Heat Entering From Plant

$$q_0 = W \rho C_p T_N + \dot{H}_{RJ} \qquad (3\text{-}19)$$

Combining Eqs. (3-15), (3-16), (3-17), and (3-19) yields the expression for the temperature of the first segment:

$$\frac{\Delta T_1}{\Delta t} = \frac{WN}{A} \frac{(T_N - T_1)}{\Delta X} - \frac{K(T_n - E)}{24 \rho C_p \Delta X} + \frac{\dot{H}_{RJ} N}{\rho C_p A \Delta X} \qquad {}^\circ F/hr \qquad (3\text{-}20)$$

4. DATA SCREENING METHODOLOGY

In this section, a method is described with which long-term weather records can be screened to find the period in which the cooling pond temperature will be maximized.

The "equilibrium temperature" heat transfer approach is used in a method that decouples the plant heat input effects from environmental effects on the pond. The temperature of the pond may be determined by the solution of the differential equation for the mixed-tank model,

$$\frac{dT}{dt} = \frac{AK}{\rho\, C_p\, V} (E - T) + \frac{\dot{H}_{RJ}}{\rho\, C_p\, V} \qquad\qquad (4\text{-}1)$$

For the purpose of developing the model, K, E, and V are temporarily assumed to be constant. Equation (4-1) will, therefore, be linear with respect to T, the fully mixed pond temperature.

Since the equation is linear, it is possible to consider that the pond temperature is a sum of the unloaded pond temperature T' and an "excess" temperature θ.

$$T = T' + \theta \qquad\qquad (4\text{-}2)$$

But T' would be determined by the solution of Eq. (4-1) without external loading:

$$\frac{dT'}{dt} = \frac{AK}{\rho\, C_p\, V} (T' - E) \qquad\qquad (4\text{-}3)$$

Subtracting Eq. (4-3) from Eq. (4-1) gives the differential equation for excess temperature

$$\frac{d\theta}{dt} = \frac{AK\theta}{\rho\, C_p\, V} + \frac{\dot{H}_{RJ}}{\rho\, C_p\, V} \qquad\qquad (4\text{-}4)$$

The determination of pond temperature has, therefore, been separated into two simpler problems, because now the ambient and excess pond temperatures can be determined independently of one another. The excess temperature θ does not depend on the meteorological record, so it can be solved directly from Eq. (4-4) using the plant heat rejection rate. The pond ambient temperature T' does not depend on the heat rejection from the plant, so it can be

21

calculated from Eq. (4-3) using only the long meteorological record. The peak pond temperature can, therefore, be found by summing (superimposing) the peak T' and θ:

$$(T)_{peak} = (T')_{peak} + \theta_{peak} \tag{4-5}$$

Unfortunately, the basic premise that Eq. (4-1) is linear is incorrect. Both K and E are functions of T. In addition, the pond volume V will change as water on the pond is lost by seepage and evaporation. (Makeup water is assumed to be unavailable during the operation of the pond.) The thermal hydraulics of the pond and, therefore, how it responds to heat input, will depend on how and when heat is rejected to the pond. If the pond can be represented by the completely mixed model, the superposition of T' and θ may overestimate the peak pond temperature for very high loadings.

The utility of the methods just described is to identify the timing of maximum ambient pond temperature T' and maximum excess temperature θ so that more accurate computations can be made in which the pond temperature T can be determined directly. A more sophisticated model of the pond may in fact be desirable for the actual pond temperature calculations rather than the simpler models employed to screen the meteorological data. The initial temperature and starting time for this computation is determined from the screening procedure. Since the heat transfer relationships are nonlinear with respect to pond temperature, and since the model ultimately used for temperature calculations may be different from those used in the screening, there are no firm guarantees that the optimal starting time for peak temperature will necessarily be found. Most likely, the optimal starting time will fall within hours of that determined by the screening procedure. A series of sensitivity runs spaced several hours apart, starting both before and after the starting time indicated by the screening procedure will assure that the peak pond temperature has indeed been found.

4.1 Meteorological Inputs to Screening Model

The screening model developed in Section 4 required two types of data: (1) weather data (dry bulb temperature, dew point, windspeed, and cloud cover) which may be obtained from National Weather Service records, and (2) rates of net solar radiation which do not exist for long periods of record. A method for synthesizing solar radiation using cloud cover data has been developed. National Weather Service tapes of Tape Data Family-14 (TDF-14) are used by the model as a source of temperature and windspeed data and the cloud cover observations. These tapes are available for major observation points throughout the United States.

The solar radiation term for the heat exchange relation must be either taken from direct measurements or estimated. The model estimates hourly solar radiation rates in a three-step process. First, given the latitude of the pond and the time of year, the maximum solar radiation available to the pond for the day under conditions is estimated. Second, this gross figure is fitted to a sinusoidal function to find the rate of insolation for each hour of daylight. Finally, these hourly rates are modified to take into account the effect of cloud cover.

A subroutine based on the work of R. W. Hamon (Ref. 8) is used to estimate the maximum daily solar radiation. This total daily radiation figure is fitted to a sinusoidal function as shown in Figure 4.1. The hourly variation of radiation is

$$\dot{H}_S(t_0) = 2t_1[\alpha\cos(\omega t_0) - \alpha\cos(\omega t_1)] \qquad Btu/(ft^2 \ day) \qquad (4\text{-}6)$$

where

$$\alpha = \frac{A_0}{\frac{1}{\omega} \sin(\omega t_1) - t_1 \cos(\omega t_1)} \qquad (4\text{-}7)$$

$$\omega = \frac{\pi}{12} \qquad hr \qquad (4\text{-}8)$$

and

A_0 = one-half the daily insolation

t_0 = time of observation before or after midday, hr

t_1 = half-length the time of daylight, hr

Solar radiation ultimately reaching the earth's surface is greatly affected by atmospheric conditions, especially by cloud cover. The amount of cloud cover in tenths of the total sky obscured is available from the data tapes. This information is used in a relationship developed by Wunderlich (Ref. 9) to modify the insolation rates:

$$\dot{H}_{SN} = \dot{H}_S(1 - 0.65 \ C^2)0.94 \qquad Btu/(ft^2 \ day) \qquad (4\text{-}9)$$

In which:

$\dot{H}_{SN}$ = net solar radiation Btu/(ft^2 day)

$\dot{H}_S$ = gross rate of solar radiation Btu/(ft^2 day)

C = cloud cover in tenths

0.94 = factor that adjusts for the average 6% reflection from the water surface

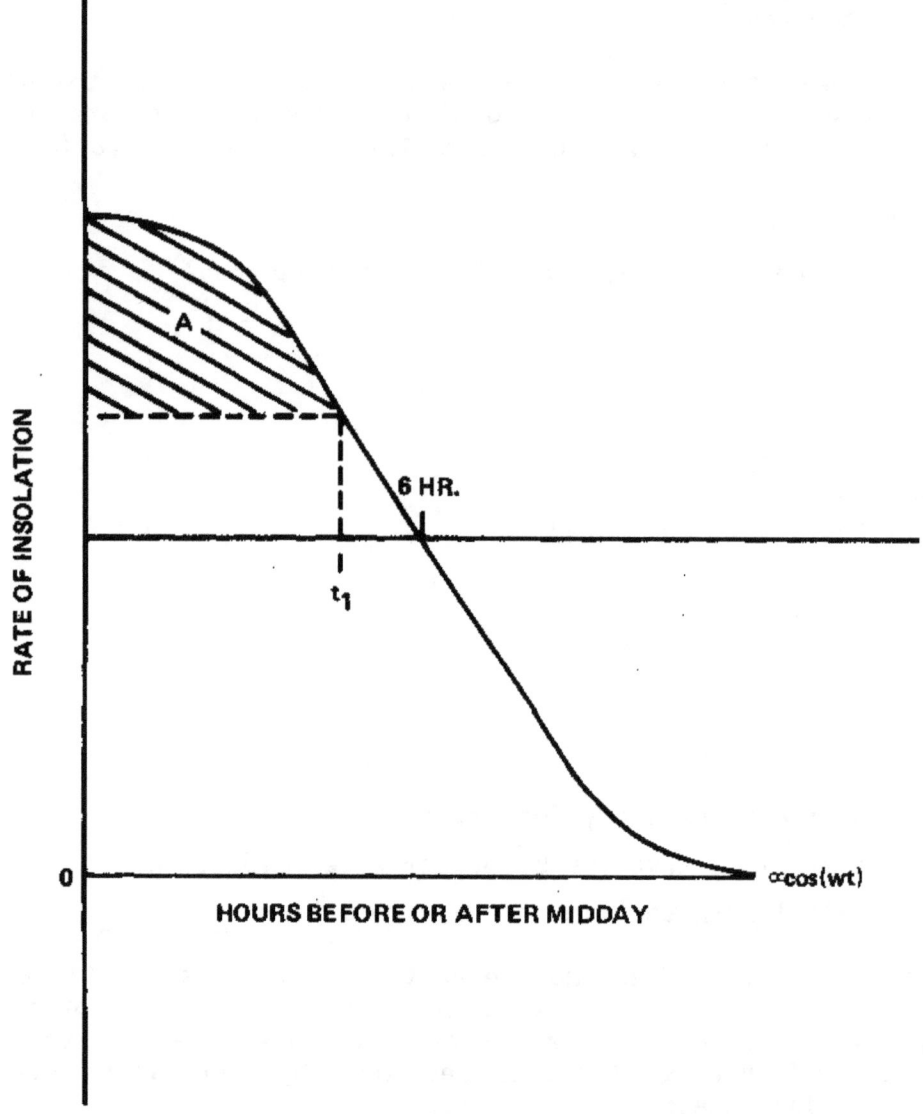

Figure 4.1 Insolation as a Function of Time.

24

5. APPLICABILITY OF WEATHER RECORD

Long-term meteorological records at the site itself are not usually available. Current NRC practice requires only limited onsite data collection. Meteorological data collected onsite may be inadequate for cooling pond analysis because measurements of solar radiation and cloud cover are not required by current regulations or guidelines.

In the absence of long-term onsite data, the meteorological data for analyzing UHS performance must be obtained from offsite weather stations (such as airports) for which long-term records, including solar radiation or cloud cover, are available. Additionally, the site and offsite data may display significant differences because of orographic effects. Because long-term records are absolutely necessary, some method to ensure applicability of the offsite data is required.

To develop the method, several assumptions are required:

(1) A representative norm of the true pond temperature will be the equilibrium temperature E, which can be calculated from the monthly mean values of the meteorological variables, and

(2) The response of the pond hydrodynamics will be fast compared to meteorological changes.

The validity of the assumptions will be subsequently shown by way of an illustrative example.

The assumptions greatly simplify the analysis of the problem and allow a meaningful quantitative appraisal of the onsite and offsite data without a dynamic analysis of the pond itself.

The principal tool in the methodology is the equilibrium temperature calculated by the transient mixed-pond model (UHS3) using monthly average data. The equilibrium temperatures are determined using offsite data [that is, $E(\bar{x})_{offsite}$] and again using onsite data [that is, $E(\bar{x})_{onsite}$]. Of course, the data cover the same period.* The difference,

$$\Delta T = E(\bar{x})_{offsite} - E(\bar{x})_{onsite},$$

is ultimately added to the peak temperature T_{max} calculated by the UHS3 model to reflect the bias induced by using the available long-term offsite data verses the onsite data.

The ΔT thus calculated represents the total bias induced by the different data sets. It may be of interest to know the relative effects of each meteorological parameter on the ΔT.

The important differences are long term. We assume that by using monthly (30-day) averages, the effects of short-term, local variations have been adequately included without having to deal analytically with such phenomena as thunderstorms.

The difference of the pond temperature ΔE, in response to difference in meteorological data onsite and offsite, can be determined by considering the partial differentials of E with respect to the independent variables T_D, T_A, U, and $\dot{H}_{SN}$:

$$dE = \frac{\partial E}{\partial T_D} dT_D + \frac{\partial E}{\partial T_A} dT_A + \frac{\partial E}{\partial U} dU + \frac{\partial E}{\partial \dot{H}_{SN}} d\dot{H}_{SN} \qquad (5\text{-}1)$$

If the differences between offsite and onsite parameters are small, the difference ΔE can be approximated by the sum of the individual differences due to changes in a single meteorological parameter, holding the others constant:

$$\Delta E \sim \Delta E_{T_A, U, \dot{H}_{SN}} + \Delta E_{T_D, U, \dot{H}_{SN}} + \Delta E_{T_A, T_D, \dot{H}_{SN}} + \Delta E_{T_A, T_D, U} \qquad (5\text{-}2)$$

The sum of the four terms on the right-hand side of Eq. (5-2) will probably not add up to the ΔE predicted directly from the calculation of E_1 and E_2 because of nonlinearities. The breakdown into individual components should be largely indicative of the true differences between the data sets, however.

A brief computer program, COMET (COmpare METeorology), has been written which evaluates the differences in steady state temperatures between two data sets and their sensitivity to differences in the averages of dew point, air temperature, windspeed, and solar radiation between the two sets of data. This program also calculates the correction factor, in cubic feet of water, for the differences in evaporation between two sites based on the 30-day average meteorology. Resultant steady state temperatures and water loss rates between the two data sets are correlated. The standard errors σ and coefficients of determination r^2 are calculated for temperature and evaporation.

The use of the correction factors from program COMET alleviates the ambiguity of location of the weather station instruments at either site. The conservatism of using the difference in equilibrium temperatures E(x) from monthly average data to correct for peak temperature at the site can be demonstrated by example. Consider that the "estimated" correction factor $\Delta T_1 = E(x)_{offsite} - E(x)_{onsite}$ is to be added to the peak temperature calculated by the UHS3 model of pond performance. A 40-day continuous record of meteorological data at Harrisburg, Pennsylvania, was used as the data base for the offsite location. To represent the onsite data base, the same 40-day record was used, but a bias was added to each of the values of T_A, T_D, H_S, and U, one parameter at a time. Peak thermally loaded pond temperatures were then calculated for the onsite data base with no bias, and recalculated with a bias on each meteorlogical parameter. The "true" correction factor is observed from this calculation as the difference in the peak temperature of the biased and unbiased case.

The true correction factor for peak temperature using the mixed-tank model were compared with the differences in equilibrium temperature based on a 40-day average of the meteorological parameters (estimated correction factor). The results of this comparison are presented in Table 5.1. The table shows

Table 5.1 Estimated Correction Factor $w\bar{E}(\bar{x})$ Compared to Actual Transient Model Response ΔT_{max}* (True Correction Factors)

Bias on meteorological terms*					$\Delta\bar{E}(x)$** (estimated correction factor) from Eq. (5-2)		ΔT_{max} true correction factors) from program UHS3
ΔT_A,*** °F	ΔT_D,+ °F	$\Delta \hat{H}_S$,++ Btu/ft²hr	ΔU,+++ mph	$\bar{E}(\bar{x})$		T_{max}	
0	0	0	0	74.93	0	104.36	0
10	0	0	0	77.44	2.51	106.01	1.65
0	10	0	0	80.89	5.96	109.14	5.03
0	0	1000	0	82.00	7.07	110.14	5.78
0	0	0	5	72.49	-2.45	97.6	-6.76

* ΔT_{max} is based on actual peak temperature with example heat load in example pond using mixed-tank model.

** $\Delta \bar{E}(x)$ is based on 40-day average of T_A, T_D, H_S, and rms U.

*** ΔT_A = bias added to each hourly value of dry bulb temperature T_A.

\+ ΔT_D = bias added to each hourly value of dew point temperature T_D.

\+\+ $\Delta \hat{H}_S$ = bias added to each hourly value of solar radiation $\hat{H}_S$.

\+\+\+ ΔU = bias added to each hourly value of windspeed U.

that, as expected, the estimated correction factors $\Delta\bar{E}(\bar{x})$ were always larger positively than the actual increases in pond temperature predicted by the mixed-tank transient model ΔT_{max}, (the "true" correction factor) and are, therefore, conservative.

Table 5.1 provides, in part, an indication of the bias induced by an arbitrary variation of each parameter. The variations $[\Delta\bar{E}(x)]$ indicate that, with the exception of solar radiation, H_S* the most important factor is the dew point.

---

*Unfortunately, the relative magnitude of the arbitrary variations (i.e., ΔT_A of 10°F is well within the range of variation potentially expected between sites: the variation of ΔH_S by 1000 Btu/ft² hr) is a major change.

6. DESCRIPTION OF COMPUTER PROGRAMS AND THEIR OPERATION

6.1 Introduction

Three separate computer programs are described which may be used for several facets of the cooling pond analysis. The programs rely on the procedures and methods described in the previous sections. All programs are written in CDC 7600 FORTRAN IV. Minor modifications may be necessary for other computer systems.

(1) Program UHSPND is used to scan the weather record tapes to predict the likely periods of lowest cooling performance and highest evaporative loss.

(2) Program COMET compares the limited quantity of onsite meteorological data with summaries of offsite data provided by program UHSPND to determine if there are significant differences between the two which might lead to differences in predicted pond performance.

(3) Program UHS3 can be used to calculate the most pessimistic cooling pond temperature using idealized pond hydraulic models and the abbreviated weather record furnished from program UHSPND. These programs are described in greater detail in the following sections.

6.2 Meteorological Data Screening Program UHSPND

Program UHSPND can be used to scan long weather records to determine the period of lowest cooling performance and highest evaporation for small cooling ponds in UHS service. A simple mixed-tank hydraulic model and the Brady-Geyer heat transfer formulae are employed in a running simulation for the entire length of the weather record. The time of maximum ambient pond temperature and the 30-day period giving maximum evaporation are determined from the simulation.

6.2.1 Program Operation

The program first reads and screens meteorological data from National Weather Service Tape Data Family 14 (TDF-14) magnetic tapes. Hourly or three-hourly values of up to 48 meteorological variables are stored on these tapes in a compact alphanumeric code. Subroutine SUB1 interprets the code and extracts the values of windspeed, dry bulb temperature, dew point temperature, cloud cover, relative humidity, and atmospheric pressure. As a computational expedient, only the months May through September are scanned, since it is highly unlikely that either the peak temperature or evaporation losses would occur in the other months.

The stored data are checked for missing or inconsistent values. If one or two consecutive observations of a meteorological parameter are missing, they will be replaced by interpolated values. If, however, more than two consecutive observations are missing or in error, the entire day of data is skipped and an message to this effect is printed.

29

The program synthesizes solar radiation needed for subsequent calculations from the cloud cover, date, and latitude, since no direct observations of solar radiation are contained in the TDF-14 tapes. This procedure is discussed in Section 4. Direct observations of solar radiation would be most desirable if available from other sources, but no provisions for their input are presently incorporated in the program.

Subroutine SUB2 numerically calculates the ambient unloaded pond temperature and evaporative loss with the mixed-tank model and the Brady-Geyer heat transfer relationships using the meteorological variables generated in subroutine SUB1. The yearly maximum pond temperature and yearly maximum 30-day evaporative water loss are determined along with their dates of occurrence.

Subroutine SUB5 statistically treats the data base consisting of the annual maximum pond temperatures and maximum annual 30-day evaporations for further manual analysis.

6.2.2 Program Outputs

The program provides the following information, depending in some cases on the options selected:

(1) An informative message is printed if missing or inconsistent data are encountered, so that it is clear that the record for that day has been skipped.

(2) A table of hourly values of windspeed, dry bulb temperature, dew point temperature, solar radiation, cloud cover, and relative humidity is printed and/or punched for the 35 days preceding the time of maximum ambient temperature and the 5 days following. This table may subsequently be used in a more rigorous computation of thermally loaded pond temperature with program UHS3, or may be used with some other dynamic temperature model. Although only the values of windspeed, dry bulb temperature, dew point temperature and solar radiation are used in the UHSPND and UHS3 models, other formulations of the heat transfer relationships may require the cloud cover and relative humidity, so these are also outputted.

(3) The dates and quantity of evaporation for the yearly worst 30-day ambient period in an unloaded pond is outputted. The quantity of (unloaded) evaporated water loss may be added to a conservative estimate of excess evaporative loss from added heat to determine total evaporative water loss directly, without the need for an additional computer program as is necessary with the peak temperature calculations.

(4) Monthly averages of meteorological parameters for all specified years of the record are printed for the purpose of comparing offsite data with limited quantities of onsite data using program COMET described later.

(5) The maximum annual ambient pond temperature and 30-day evaporation for all years on the tape are printed, ranked in order from highest to lowest magnitude. Approximate probabilities are calculated so that the ranked outputs can be plotted on an arithmetic-probability scale. The mean,

standard deviation, and skew of the data are also printed. Further statistical manipulation may be performed manually using the procedures outlined in Appendix A, Statistical Treatment of Output.

6.2.3 Program Inputs

The following input data are necessary to run program UHSPND:

(1) Pond surface area, ft^2 or acres.

(2) Pond volume, ft^3.

(3) Latitude, °N.

(4) A TDF-14 weather tape from a representative station near the site.

The TDF-14 weather tapes can be obtained from the National Climatic Center, Federal Building, Asheville, North Carolina 28801.

Computer and peripheral requirements to run program UHSPND on the Brookhaven National Laboratories CDC 7600 computer are one magnetic tape drive, two disk files, and about 12,000 (decimal) words.

Specific instructions for running the program apply to the NRC version on the Brookhaven computer. Versions of UHSPND for use at other centers can be expected to differ in their use of tapes and job control cards. Minor modifications to the program may be necessary for computer systems other than CDC.

The data deck required to operate program UHSPND consists of three types of data cards; the pond data card, the monthly average card, and the end card. The input data are read in NAMELIST form named INPUT.

The following tables explain the meaning of each variables in the NAMELIST:

6.2.3.1 Pond Data Card

This NAMELIST specifies the pond parameters for the mixed-tank models and specifies certain printing options as shown in Table 6.1.

6.2.3.2 Monthly Average Card

This NAMELIST specifies the year and month to start computing monthly meteorological summaries to be used for comparison with onsite meteorological data.

6.2.3.3 End Card

By specifying N = 0, the program terminates.

One set of output is generated from each pond data card or monthly average card. These cards are unrelated and may be inserted in any order.

Table 6.1 Pond Data Card Input Parameters

Variable	Value	Type and description
N	1-99	Integer--card number used to identify the the card as a "pond data" card and to identify the results in the output
A	≥ 0	Real, pond surface area in square feet
	<0	Real, pond surface area in acres
V	≥ 0	Real, pond volume in cubic feet
	<0	Real, pond volume in acre-feet
LAT	25-50	Real, latitude of pond in decimal degrees north latitude
IPRNT		Integer--print option
	0	Prints and punches hourly meteorological data
	1	Printed output only
	-1	Punched output only

If a second pond data or monthly average card is used, say to test the sensitivity to a variation in a pond parameter, only the variable changed needs to be inputted on the NAMELIST card.

6.2.4 Data Input Example

Consider a pond with volume 1.5×10^7 ft^3 and 8×10^5 ft^2 surface area at latitude 45°N. Determine the highest ambient temperature and evaporation rate, and print the worst case meteorology. Rerun the calculation with half the volume, determine the periods of highest temperature and evaporation, and print and punch the output for the worst temperature period. Finally, compute the monthly averages of the meteorological data from June 1971 to the end of the tape.

The data input for this example would be:

```
$INPUT N=1, A=8.0E5, V=1.5E7, LAT=45.0, PRINT=1$
$INPUT N=2, V=0.75E7, ISRCH=1, IPRINT=0$
$INPUT N=101, YRMODY(1)=71, YRMODY(2)=6$
$INPUT N=0$
```

Table 6.2 Monthly Average Card Input Parameters

Variable	Value	Type and description
N	> 99	Integer--identified this card as a monthly average card.
YRMODY(1)		Real, the year of the beginning date for the computation of monthly averages of meteorological data
YRMODY(2)	5-9	Real, the month of the beginning date for the computation of monthly averages.
LAT	25-50	Real, the latitude in decimal degrees north if different from that previously specified.

6.3 Program COMET

Program COMET (COmpare METeorology) compares equilibrium temperature and evaporation rates computed from monthly average values of solar radiation, dew point temperature, dry bulb temperature, and rms (root mean square) windspeed for two data sets. It has been previously demonstrated in Section 5 that equilibrium temperature computed from monthly average meteorological conditions can be a meaningful norm for the comparison of two data sets used to compute peak temperatures.

Program UHSPND computes the monthly averages of the meteorological parameters from the offsite weather station record provided on the National Climatic Center tape. The other data set would be taken from limited onsite measurements.

If onsite data are not complete (for example, if solar radiation is not available), the offsite data can be substituted for the missing parameters. The program calculates the equilibrium temperature $E(x)$ and 30-day evaporation $W_e(x)$ for each data set, the difference in calculated values of E, and the apparent differences in E due to differences between each of the meteorological parameters. Therefore, if one of the meteorological parameters for the site is unknown, the apparent differences due to only the other three parameters can still be determined.

The output values of onsite and offsite equilibrium temperature and evaporation rates are correlated for as many months as available to determine if there is a significant difference between the locations. The coefficient of determination r^2 is computed for $E(x)$ onsite and offsite. A coefficient of determination of 0.9 would indicate that 90% of the variance in one data set is accounted for by variation of the other data set, and that 10% of the variation is unexplained.

The average equilibrium temperature difference and average evaporation rate difference between the two data sets are the biases $E(\bar{x})$ and $W_e(\bar{x})$, respectively. The biases may be used cautiously as correction factors to the peak loaded pond temperature and 30-day evaporation loss. The coefficient of determination r^2 should be high. Lower values may indicate poor quality data, real orographic differences between sites, or a combination of the two. Because the data bases are generally small and may be incomplete, we suggest that the biases be used only in the conservative sense; that is, if onsite $E(\bar{x})$ or $W_e(\bar{x})$ are greater than corresponding offsite values, the difference should be added to the peak loaded pond temperature or evaporation as a correction. If the opposite is the case, no corrections should be made.

6.3.1 Program Inputs

Program COMET requires monthly averages of dry bulb temperature, dew point temperature, solar radiation and rms windspeed for each site. The first card specifies the number of months of data I and is read in I5 format. The next I cards contain the following information read in 8F10.0 format:

Field	Variable	Description
1	TD1	Dew point temperature, °F, data set 1
2	TA1	Dry bulb temperature, °F, data set 1
3	W1	Rms windspeed, mph, data set 1
4	H1	Solar radiation, Btu/(ft^2 day), data set 1
5	TD2	Dew point temperature, °F, data set 2
6	TA2	Dry bulb temperature, °F, data set 2
7	W2	Rms wind speed, mph, data set 2
8	H2	Solar radiation, Btu/(ft^2 day) data set 2

If dew point temperature is not available directly, it can be synthesized from dry bulb temperature, wet bulb temperature, and atmospheric pressure using a psychrometric chart, or subroutine PSY1 described in Appendix B.

6.4 Program UHS3

Program UHS3 calculates the temperature in the ultimate heat sink pond under the combined influence of the meteorology and the external plant heat load. Hourly meteorological data are provided on cards from program UHSPND.

The pond is represented by three simplified hydraulic models simultaneously: the mixed-tank model as used in the screening program UHSPND, the stratified-flow model, and the plug-flow model. Heat transfer relationships are based on the Brady-Geyer method as in program UHSPND.

The pond outlet temperatures and volume for all three models are printed simultaneously. Maximum temperature for each pond model is determined and the time of occurrence of the maximum is printed.

6.4.1 Program Input

Necessary input data for this program include a title card, the external heat input, meteorological conditions, volume and surface area, makeup, blowdown, leakage, and circulation flowrate of the pond:

(1) The first card of the data deck is a title card. Information entered on this card will be printed at the beginning of the program output. If no information is to be printed out, this card should be left blank.

(2) Meteorological data are generally provided directly from program UHSPND. The first card in the meteorological deck specifies the number of time periods in the table (usually 960) and is read in I5 format. The subsequent cards are read two time periods (usually 1 hour each) per card as illustrated in Table 6.3.

   If constant meteorological conditions are specified, only the first values of W, TA, TD, and HSN need to be inputted.

(3) The heat and flowrate table is inputted next. The plant heat rejection and UHS flowrate during the design accident should be plotted on a semilog plot, with heat and flowrate on the linear scale and time on the logarithmic scale. A table of heat and flowrate to the pond versus time should then be created from a straight line approximation of the graph. This procedure must be followed because a log-linear interpolation of the heat and flowrate table is used in the program. Plant heat is often provided in this graphic form directly.

   Heat and flowrate are inputted in a NAMELIST format named HFT. For example, a typical heat load and flowrate table would be:

   $HFT HEAT(1) = 0.0, 1.0E8, 1.6E8, 1.0E8, 4.9E8, 5.6E8, 6.9E8, 5.1E8, 1.4E8, 0.8E8, 0.5E8, FLOW(1) = 40, 50, 8*60, TH(1) = 0.001, 0.025, 0.04, 0.08, 6, 50, 600, 1000, NH = 9$

   where

   HEAT = array of plant heat input, Btu/hr

   FLOW = array of flow inputs, ft³/hr

   TH = array of corresponding times (relative to the start of the accident) for the HEAT and FLOW arrays

   NH = number of entries in the table.

Table 6.3 Meteorological Input for Program UHS3
Format [13, 2(3F5.1, F6.1, F4.0)]

Field	Variable	Description
1	ISEQ	Sequence number--not used
2	W(I)	Windspeed, mph
3	TA(I)	Dry bulb temperature, °F
4	TD(I)	Dew point temperature, °F
5	HS(I)	Solar radiation, Btu/(ft² day)
6	CC	Cloud cover--not used in this program but punched from UHSPND
7	RH	Relative humidity--not used in this program, but punched from UHSPND
8	W(I+1)	Windspeed--second set on card
9	TA(I+1)	Dry bulb temperature, °F
10	TD(I+1)	Dew point temperature, °F
11	HS(I+1)	Solar radiation
12	CC	Cloud cover
13	RH	Relative humidity

It should be noted that the start of the heat and flowrate table does not necessarily have to correspond to the start of the meteorological input table. The time for the start of the heat and flowrate table is delayed by a variable TSKIP (HR), described below.

(4) Pond parameters and constants are read next in a NAMELIST format called INLIST. The variables in INLIST are described in Table 6.4.

6.4.2 Utilization of Program UHS3

Program UHS3 is usually employed to determine maximum pond temperature in the following manner:

(1) Two initial pond simulations should be performed (in the same run):

(a) The first run simulates the pond ambient temperature resulting only from meteorological inputs without the external heat load. This is most easily done by setting TSKIP to a large number of hours in INLIST (for example, TSKIP=5000). The peak ambient pond temperature and time of occurrence generally will not be the same as those predicted from UHSPND.

36

Table 6.4 NAMELIST INLIST for Program UHS3

Variable	Default value	Description
VZERO	0.0	Pond volumes, ft^3--if zero, terminates program
BLOW	0.0	Blowdown flow out, ft^3/hr
A	0.0	Pond surface area, ft^2
NSTEPS	100	Number of timesteps to be performed
NPRINT	10	Printouts of pond temperatures every NPRINT steps
DT	0.2	Integration timestep, hours
TZERO	80	Initial pond temperature, °F
TSKIP	0	Time after start of program that corresponds to start of heat and flow table. Shifts this table relative to meteorology table which starts at time zero. For time less than TSKIP, evaporation is suppressed so that the pond volume does not decrease.
QBASE	0	Bias to be added to all HEAT in heat-flow table, Btu/hr
FBASE	0	Bias to be added to all flowrate in heat-flow table, ft^3/hr
E	80	Constant equilibrium temperature °F, if so specified by IMET=1
AK1	150	Constant surface heat exchange coefficient, $Btu/(ft^2\ day)/°F$ if IMET=1
IMET	0	Optional constant E and AK1 if IMET=1
BTA	0	Bias to be added to all TA in table (dry bulb temperature), °F
BTD	0	Bias to be added to all TD in table (dew point temperature), °F
BHS	0	Bias to be added to all HS (solar radiation), $Btu/(ft^2\ hr)$
BW	0	Bias to be called to all W in Table 6.3 (windspeed), mph
HEAT FLOW NH	Same as specified input in NAMELIST HFT	Heat-flow table if different from that specified by previous input in NAMELIST HFT

Note: Multiple runs may be made by inserting several INLIST cards in succession. Only the variables which are different from the previous namelist card read are changed. The program terminates by setting VZERO=0.

(b) The second simulation determines the peak pond temperature only from the effects of external heat input. This is done by resetting TSKIP to zero, and specifying that E and AK1 are constants in namelist INLIST (for example, IMET=1, TSKIP=0, AK1=120, E=85, TZERO=85). The choice of AK1 and E is somewhat arbitrary, but the initial pond temperature TZERO should always be set equal to E.

(2) A second run is prepared so that peak ambient pond temperature determined from the first simulation will coincide with the peak excess temperature caused by plant input alone:

(a) By inspection of the two previous simulations, choose the model desired (for example, mixed tank) and the time of peak temperature for each.

(b) The approximate time to delay the start of the heat input TSKIP is then defined:

TSKIP = time of peak ambient temperature minus time of peak excess temperature.

(d) Because of nonlinearities in the pond models, the peak temperature will not necessarily concide with that of the direct linear superposition, and the time to the peak may be shifted. Several simulations may be made within the same run, varying the parameters TSKIP by several hours to assure that the peak temperature has been found, although in general the differences should be minor.

An example run of all programs from start to finish will be covered in the next section.

7. SAMPLE CALCULATIONS

This section describes the analysis of a hypothetical UHS cooling pond and shows how the computer programs and methods presented in this paper can be used with historical weather records to determine the design basis return temperature and worst-case 30-day evaporative water loss for a given pond.

The following information is needed for computer programs UHSPND, COMET, and UHS3 in order to perform these analyses.

(1) For UHSPND:

    (a) Pond area and volume.

    (b) Latitude of the pond.

    (c) A National Weather Service data tape (TDF-14) for an observation point near the pond site.

    (d) Date of the beginning of onsite data collection.

(2) For COMET:

    (a) Monthly averages for the months of May through September of onsite observations of daily insolation, dry bulb temperature, dew point temperature; and the monthly rms windspeed. If onsite insolation is not available, the offsite insolation term generated by UHSPND can be used in its place as long as this fact is acknowledged in the analysis of the results.

    (b) Monthly averages as described above from the long-term (offsite) weather record for the period that corresponds to the period of onsite meteorological observations. This information can be obtained from UHSPND by using a monthly average card in the data deck.

(3) For UHS3:

    (a) The punched output from UHSPND consisting of the meteorological parameters for the 40-day period that encompasses peak ambient pond temperature.

    (b) A heat-flowrate table describing the heat rejected by the plant and the flowrate through the pond during the period of time following a design basis accident.

    (c) Pond initial volume and surface area.

    (d) Blowdown and seepage rates for the pond.

7.1 Finding the Period of Worst-Case Cooling Performance and 30-Day Evaporative Water Loss--Program UHSPND

The first step in the analysis is to use UHSPND to find the periods of recorded weather data that will result in the worst-case cooling performance (that is, highest pond temperature) and highest 30-day evaporative water loss. UHSPND can also be used at this point to generate the monthly averages of the meteorologic parameters needed to run program COMET.

A hypothetical pond located at 40.25°N and having a surface area of 40 acres (1,742,400 ft²) and a volume of 320 acre-feet (13,939,200 ft³) is used in this sample analysis. The long-term (1948-75) weather record from Harrisburg, Pennsylvania, is used. Limited onsite data from a facility located on the Susquehanna River are used as the onsite record for the hypothetical pond.

The data deck for UHSPND has been constructed as described in Section 6. The period of onsite data available at the time of this study was January 1, 1973, to December 31, 1976. Since UHSPND only scans the weather record during the months of May through September, the first month of the long-term record for which onsite data are available is May 1973. Entering this month and year on a monthly average card in the UHSPND data deck causes the program to print the monthly averages necessary to run COMET for each summer month, beginning with May 1973, until the end of the long-term weather record is reached. Figure 7.1 shows the data deck for UHSPND used for this example.

The following information is printed by UHSPND as a result of the data supplied in the data deck:

(1) A list of the dates ignored by UHSPND due to periods of bad data in the long-term record.

(2) A list of the pond parameters used by UHSPND to run its pond model.

(3) A table of the yearly maximum modeled pond temperatures and 30-day evaporative losses, their dates of occurrence and their "plotting positions." Both the temperatures and evaporative losses have been ranked from highest to lowest magnitude and their sample means, standard deviations, and skews have been calculated.

(4) The daily meteorological data consisting of hourly observations for each day in the period of the 35 days ending with the date of the highest modeled pond temperature and 5 days following it. This information is also punched on cards as a result of using IPRNT=0 and a message indicating the number of the cards punched follows the printed output. No printed or punched output is provided for the days skipped because of bad data.

(5) A table of the monthly rms windspeeds and mean values of dry bulb temperature, dewpoint temperature, daily solar radiation, cloud cover in tenths, and relative humidity for the months of May through September during the period beginning May 1973 and continuing to the end of the long-term record in 1975.

```
$INPUT N=1,A=1742400.,V=13939200.,LAT=40.25,ISRCH=1,IPRNT=0$
$INPUT N=100,YRMODY(1)=73,YRMODY(2)=5,LAT=40.25$
$INPUT N=0$
```

Figure 7.1 Listing of Input for Program ULTSNK.

A partial listing of the output generated by UHSPND in this example is provided in Figure 7.2.

7.2 Statistical Treatment of UHSPND Output

The statistical methods of frequency analysis using Pearson type III coordinates, outlined in Appendix A, have been applied to the sample of yearly maximum pond temperatures in order to gain some insight into the trend in the data. The histogram in Figure 7.3 gives some idea of the distribution of the yearly maximum temperatures. A frequency plot of the yearly maximum temperature data is presented in Figure 7.4. Here the temperatures were first plotted on arithmetic-probability paper using the exceedence frequencies (plotting positions) computed by UHSPND. Next, the most likely probability curve and the 5% and 95% error bands were constructed from the mean and standard deviation computed by UHSPND and the methods and tables of Appendix A.

Note that the skew was taken to be zero because of the small size of the sample. The computed frequency curve can be used to extrapolate the 1% per year ambient exceedence pond temperature from the UHSPND results. This temperature is found to be 85.5° F. Since this is less than the maximum modeled temperature of 85.7° F, no temperature correction factor will be used in subsequent calculations. Note, however, that the maximum falls within the 5% and 95% confidence limits and is not considered anomalous.

These statistical procedures can also be applied to the sample of yearly maximum 30-day evaporative loss. The predicted loss is 992,000 ft^3. Again, the modeled maximum evaporative loss of 1,023,650 ft^3 is larger than the 1% per year exeedence loss found by extrapolation and no correction will be made for this in subsequent evaporation calculations.

7.3 Determining the Applicability of the Offsite Data Set--Program COMET

There is a potential for error because of the use of an offsite data record. The second step in the cooling pond analysis is to compare the offsite record with the limited onsite record and generate some reasonable correction factors if a significant difference exists between the two records. Program COMET is used for this task.

COMET compares equilibrium temperatures generated from the monthly arithmetic mean values of dew point, dry bulb temperatures, daily solar radiation, and the root mean square (rms) windspeeds from two different sites. The rms windspeed is used as the representative average because of the quadratic function of windspeed in the model equations. This information is input to COMET in the form described in Section 6, one month per data card. In this case, a period of 15 spring and summer months, May 1973-September 1975, was available for study. The offsite information was input as set 1 and the onsite information

```
U.S. NUCLEAR REGULATORY COMMISSION- ULTIMATE HEAT SINK COOLING POND METEOROLOGICAL SCANNING MODEL
R CODELL AND W NUTTLE, NOVEMBER 1979

********** SUBROUTINE SUB1 HAS BEEN CALLED FOR LATITUDE = 40.25 DEG. NORTH *****

                DISCONTINUITY IN DATA CAUSED    6/11/71 TO BE SKIPPED
                DISCONTINUITY IN DATA CAUSED    9/25/71 TO BE SKIPPED
                DISCONTINUITY IN DATA CAUSED    5/ 5/72 TO BE SKIPPED
                DISCONTINUITY IN DATA CAUSED    5/ 6/72 TO BE SKIPPED
                DISCONTINUITY IN DATA CAUSED    5/ 7/72 TO BE SKIPPED
                DISCONTINUITY IN DATA CAUSED    5/ 8/72 TO BE SKIPPED
                DISCONTINUITY IN DATA CAUSED    8/ 7/72 TO BE SKIPPED
                DISCONTINUITY IN DATA CAUSED    7/11/73 TO BE SKIPPED
                DISCONTINUITY IN DATA CAUSED    7/15/73 TO BE SKIPPED
                DISCONTINUITY IN DATA CAUSED    7/16/73 TO BE SKIPPED
                DISCONTINUITY IN DATA CAUSED    7/17/73 TO BE SKIPPED
                DISCONTINUITY IN DATA CAUSED    5/ 1/75 TO BE SKIPPED


********** POND NUMBER  1 HAS THE FOLLOWING PARAMETERS ****************************

        SURFACE AREA      1742400.00 FT**2  (      40.00 ACRES)

        VOLUME          13939200.00 FT**3  (     320.00 ACRE-FT)

        ISRCH = 1              IPRNT = 0


********** POND NUMBER  1 HAS BEEN MODELLED TO DETERMINE THE WORST ************
                PERIODS FOR COOLING AND EVAPORATIVE WATER LOSS
```

Figure 7.2 Output From Program UHSPND.

42

********THE SAMPLE OF YEARLY MAXIMUM POND TEMPERATURES AND 30 DAY *********
EVAPORATIVE LOSSES GENERATED BY THIS MODEL IS DESCRIBED BELOW.

EXCEEDED	*TEMPERATURE*	*DATE*	*EXCEEDED*	*EVAPORATIVE LOSS*	*DATE*
/100 YR	*(DEG.F)*	*(YR.MO.DY.)*	*/100 YR*	*FT**3*	*(YR.MO.DY.)*
2.45	85.71	72. 7.23.	2.45	1024361.9	66. 7.23.
5.97	83.50	75. 8. 3.	5.97	948619.5	63. 7.10.
9.49	83.15	52. 7.23.	9.49	904377.5	55. 8. 9.
13.01	82.82	68. 7.19.	13.01	887364.3	57. 7.27.
16.54	82.00	49. 7.30.	16.54	870331.8	74. 7.22.
20.06	81.85	73. 7.10.	20.06	870192.1	71. 7.28.
23.58	81.29	57. 6.18.	23.58	864591.4	54. 8.12.
27.10	81.15	64. 7.20.	27.10	855142.4	61. 7.12.
30.63	80.98	63. 7. 2.	30.63	636959.1	65. 7. 9.
34.15	80.90	59. 7. 1.	34.15	836561.6	62. 7.27.
37.67	80.84	70. 8. 2.	37.67	826182.2	59. 7. 9.
41.19	80.46	71. 7. 1.	41.19	823344.4	53. 9.13.
44.72	80.44	61. 7.25.	44.72	801589.5	64. 6.17.
48.24	80.38	48. 8.29.	48.24	794705.2	73. 8. 6.
51.76	80.30	53. 7.21.	51.76	783720.5	68. 7.31.
55.28	80.25	55. 7.28.	55.28	782405.7	70. 8.31.
58.81	80.12	65. 8.18.	58.81	764858.0	60. 7. 6.
62.33	79.91	74. 7. 9.	62.33	763910.1	56. 7.15.
65.85	78.99	51. 7.17.	65.85	754682.3	72. 8.21.
69.37	78.68	66. 6.28.	69.37	753861.1	52. 7.29.
72.90	78.42	67. 6.24.	72.90	741111.9	75. 8.19.
76.42	78.23	69. 7.18.	76.42	739849.6	51. 8. 3.
79.94	78.19	50. 8. 3.	79.94	736522.0	58. 6.18.
83.46	77.92	60. 9. 1.	83.46	735030.0	67. 7. 6.
86.99	77.88	58. 7. 8.	86.99	728814.0	69. 6.19.
90.51	77.86	62. 7. 8.	90.51	719320.1	49. 6.23.
94.03	77.08	54. 7.31.	94.03	691244.7	50. 7. 3.
97.55	76.98	56. 8.18.	97.55	671605.5	48. 7.29.

MEAN   80.22   803973.5

STANDARD DEV.   2.089   80296.12

SKEW   .545   .773

Figure 7.2 (Continued).

HOUR	WIND SP. (MPH)	DRY BULB (DEG.F)	DEWPOINT (DEG.F)	SOLAR RAD (BTU/FT2/D)	CLOUD COVER	RELATIVE HUMIDITY
0.	1.9	71.3	66.3	0.0	1.00	84.0
1.	0.0	71.0	66.0	0.0	1.00	84.0
2.	1.5	71.0	66.0	0.0	1.00	84.0
3.	3.1	71.0	66.0	0.0	1.00	84.0
4.	4.6	71.0	66.0	0.0	1.00	84.0
5.	5.0	71.3	66.7	134.6	1.00	85.0
6.	5.4	71.7	67.3	406.2	1.00	86.0
7.	5.8	72.0	68.0	677.8	1.00	87.0
8.	5.8	73.7	68.0	931.0	1.00	82.7
9.	5.8	75.3	68.0	1148.3	1.00	78.3
10.	5.8	77.0	68.0	1315.1	1.00	74.0
11.	6.9	78.7	68.7	1759.8	.93	71.7
12.	8.1	80.3	69.3	2128.5	.87	69.3
13.	9.2	82.0	70.0	2369.2	.80	67.0
14.	8.8	82.3	70.0	2321.8	.77	66.3
15.	8.4	82.7	70.0	2134.0	.73	65.7
16.	8.1	83.0	70.0	1812.7	.70	65.0
17.	8.1	82.0	69.7	1259.7	.73	66.3
18.	8.1	81.0	69.3	717.2	.77	67.7
19.	8.1	80.0	69.0	224.6	.80	69.0
20.	8.1	78.0	67.3	0.0	.63	69.7
21.	8.1	76.0	65.7	0.0	.47	70.3
22.	8.1	74.0	64.0	0.0	.30	71.7
23.	7.7	73.0	64.0	0.0	.20	73.7

Figure 7.2 (Continued).

HOUR	WIND SP. (MPH)	DRY BULB (DEG.F)	DEWPOINT (DEG.F)	SOLAR RAD (BTU/FT2/D)	CLOUD COVER	RELATIVE HUMIDITY
0.	7.3	72.0	64.0	0.0	.10	76.3
1.	6.9	71.0	64.0	0.0	0.00	79.0
2.	4.6	70.3	64.3	0.0	.27	81.7
3.	2.3	69.7	64.7	0.0	.53	84.3
4.	0.0	69.0	65.0	0.0	.80	87.0
5.	3.1	70.3	65.7	213.4	.83	85.3
6.	6.1	71.7	66.3	595.8	.87	83.7
7.	9.2	73.0	67.0	918.4	.90	82.0
8.	9.2	74.3	67.3	1194.7	.93	79.3
9.	9.2	75.7	67.7	1288.7	.97	76.7
10.	9.2	77.0	68.0	1315.4	1.00	74.0
11.	10.0	78.0	67.3	1420.2	1.00	70.0
12.	10.7	79.0	66.7	1455.9	1.00	66.0
13.	11.5	80.0	66.0	1420.2	1.00	62.0
14.	10.7	79.7	66.3	1315.4	1.00	63.7
15.	10.0	79.3	66.7	1148.8	1.00	65.3
16.	9.2	79.0	67.0	931.7	1.00	67.0
17.	10.7	78.0	67.0	678.8	1.00	69.3
18.	12.3	77.0	67.0	407.5	1.00	71.7
19.	13.8	76.0	67.0	136.1	1.00	74.0
20.	12.3	75.3	67.0	0.0	1.00	75.7
21.	10.7	74.7	67.0	0.0	1.00	77.3
22.	9.2	74.0	67.0	0.0	1.00	79.0
23.	11.1	74.0	66.7	0.0	1.00	78.0

Figure 7.2 (Continued).

********** METEOROLOGY FOR   7/29/72 **********

OUTPUT FROM PROGRAM UHSPND FOR THE PERIOD
6-21-72 THROUGH 7-28-72
HAS BEEN OMITTED BECAUSE OF ITS LENGTH

HOUR	WIND SP. (MPH)	DRY BULB (DEG.F)	DEWPOINT (DEG.F)	SOLAR RAD (BTU/FT2/D.	CLOUD COVER	RELATIVE HUMIDITY
0.	0.0	61.3	56.0	0.0	.07	83.0
1.	0.0	60.0	56.0	0.0	0.00	87.0
2.	1.9	59.3	55.7	0.0	.07	88.0
3.	3.8	58.7	55.3	0.0	.13	89.0
4.	5.8	58.0	55.0	0.0	.20	90.0
5.	5.4	59.7	56.0	68.0	.33	88.0
6.	5.0	61.3	57.0	726.0	.47	86.0
7.	4.6	63.0	58.0	1239.5	.60	84.0
8.	5.0	65.0	57.3	1520.7	.73	77.0
9.	5.4	67.0	56.7	1512.7	.87	70.0
10.	5.8	69.0	56.0	1200.6	1.00	63.0
11.	5.0	71.7	56.3	1617.2	.93	58.7
12.	4.2	74.3	56.7	1960.1	.87	54.3
13.	3.5	77.0	57.0	2177.3	.80	50.0
14.	2.3	78.0	56.7	2003.2	.80	48.0
15.	1.2	79.0	56.3	1726.3	.80	46.0
16.	0.0	80.0	56.0	1365.3	.80	44.0
17.	0.0	78.3	56.3	828.1	.87	47.3
18.	0.0	76.7	56.7	366.8	.93	50.7
19.	0.0	75.0	57.0	25.7	1.00	54.0
20.	0.0	75.0	56.7	0.0	1.00	53.3
21.	0.0	75.0	56.3	0.0	1.00	52.7
22.	0.0	75.0	56.0	0.0	1.00	52.0
23.	0.0	71.7	56.7	0.0	.93	60.7

**********NUMBER OF CARDS PUNCHED = 492 **********

Figure 7.2 (Continued).

********* THE MONTHLY AVERAGE VALUES FROM 5/ 1/73 TO END OF DATA ***********

	RMS WIND SPEED	DRY BULB (DEG.F)	DEWPOINT (DEG.F)	SOLAR RADIATION	CLOUD COVER	RELATIVE HUMIDITY
1973						
MAY	8.91	57.38	47.16	1381.6	.68	72.0
JUNE	6.12	72.79	62.67	1662.5	.61	73.1
JULY	6.43	76.14	63.78	1888.2	.46	67.7
AUGUST	5.90	75.38	64.59	1549.9	.52	71.2
SEPTEMBER	7.37	67.87	55.93	1309.3	.51	68.0
1974						
MAY	8.61	63.47	46.71	1653.4	.57	56.8
JUNE	7.59	70.60	57.27	1687.0	.61	64.7
JULY	7.54	77.27	59.89	1766.7	.51	57.6
AUGUST	5.74	76.47	63.89	1386.5	.62	66.3
SEPTEMBER	7.46	64.24	55.62	1199.7	.62	75.1
1975						
MAY	6.65	64.74	55.96	1563.7	.66	76.2
JUNE	7.61	70.57	62.24	1636.6	.58	77.0
JULY	6.84	75.01	66.34	1750.1	.50	76.7
AUGUST	6.75	75.12	66.77	1517.8	.60	77.7
SEPTEMBER	7.31	62.82	56.25	1173.4	.60	81.3

Figure 7.2 (Continued).

47

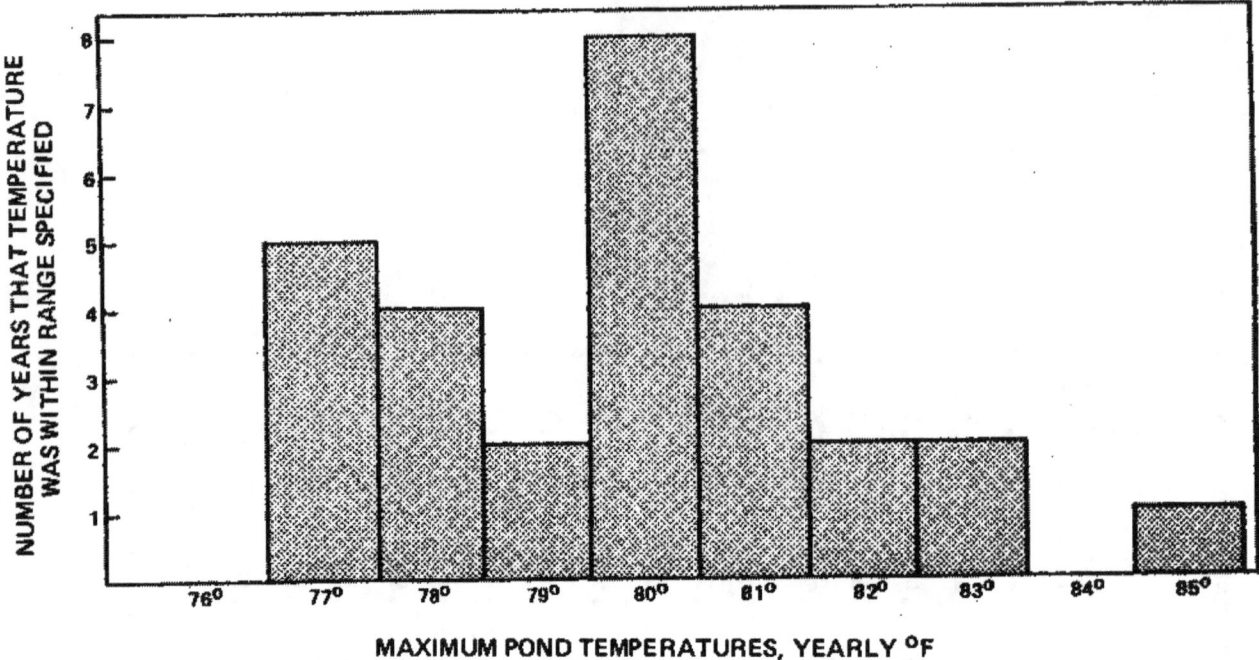

Figure 7.3 Yearly Maximum Ambient Pond Temperatures.

as set 2. The latter record lacked two of the parameters needed, solar radiation and dew point. The onsite information did include wet bulb temperature observations which allowed calculation of dew point temperatures using subroutine PSY1 presented in Appendix B. The synthesized solar radiation values from the offsite record provided by UHSPND were substituted for the onsite solar radiation. A copy of the input file for COMET is shown in Figure 7.5. Output is generated by COMET for each data card containing monthly averages. This output consists of the following information:

(1) Monthly meteorologic averages as input for each.

(2) Calculated monthly average equilibrium temperature and evaporation.

(3) Differences between the average equilibrium temperatures and evaporations of the two data sets.

(4) Component differences in the equilibrium temperature due to each meteorological parameter.

In addition to this output, the following information is printed once all of the monthly data have been read:

(1) Coefficients of determination r^2 for the equilibrium temperatures and evaporations.

(2) Biases between the two data sets for the equilibrium temperatures and evaporations.

48

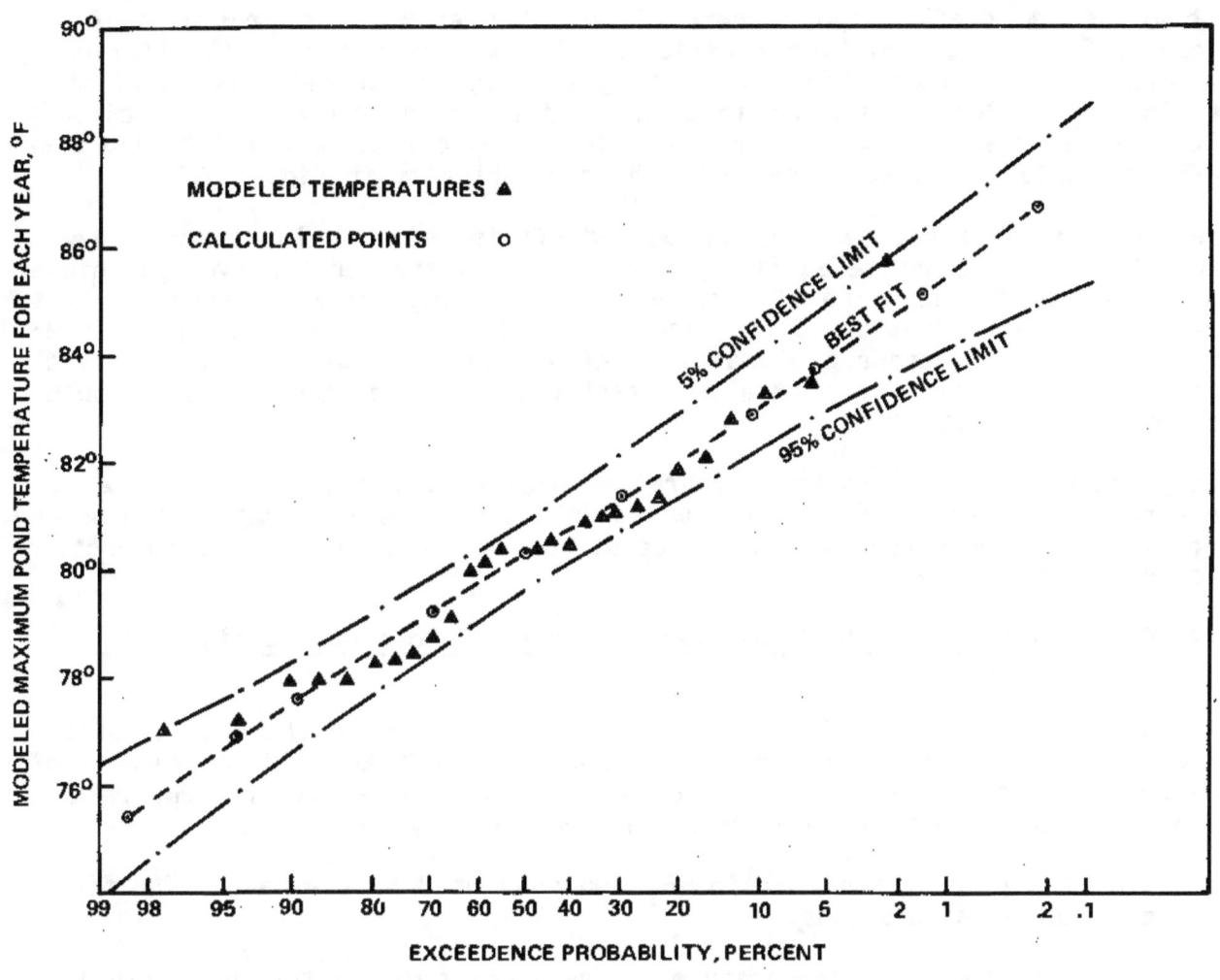

Figure 7.4 Yearly Maximum Ambient Pond Temperatures, Maximum Likelihood Frequency Curve, 0.05 and 0.95 Error Bands.

```
1 5
     47.2      57.4      8.91     1381.6      46.8      53.8      5.48
     62.7      72.8      6.12     1662.5      61.3      67.5      3.9
     63.8      76.1      6.43     1888.2      63.9      67.8      3.7
     64.6      75.4      5.9      1549.9      64.8      69.4      3.2
     55.9      67.9      7.37     1309.3      54.7      60.4      4.
     46.7      63.5      8.61     1653.4      45.       56.7      5.6
     57.3      70.6      7.59     1687.       54.5      63.1      4.8
     59.9      77.3      7.54     1766.7      59.9      68.4      4.12
     63.9      76.5      5.74     1386.5      62.4      68.       3.39
     55.6      64.2      7.46     1199.7      53.2      58.5      4.18
     56.       64.7      6.65     1563.7      53.6      61.0      4.36
     62.2      70.6      7.61     1636.6      59.9      65.7      4.53
     66.3      75.       6.84     1750.1      63.9      69.4      3.54
     66.8      75.1      6.75     1517.8      62.6      68.2      3.94
     56.3      62.8      7.31     1173.4      52.9      57.9      4.47
```

Figure 7.5 Input to Program COMET.

49

A copy of the COMET output generated for this example is presented in Figure 7.6. The correlation coefficient for the two sets of equilibrium temperatures is high (0.976), indicating that the predicted effects of the offsite meteorology on the hypothetical pond's temperature correlates closely with the effects that would have been produced by onsite meteorology had that been available. This correlation is shown graphically in Figure 7.7.

The positive bias between the onsite and offsite equilibrium temperatures indicates that the onsite equilibrium temperatures are, on the average, higher than those predicted using offsite meteorology. The primary reason for this bias is indicated from the differences due to individual meteorologic parameters. Onsite windspeeds are smaller, which leads to lower evaporation and cooling. This effect is partially offset by higher dew point and dry bulb temperatures onsite.

The positive (onsite-offsite) temperature bias will therefore be used as a temperature correction factor in subsequent calculations. Evaporation is on the average, higher for the offsite data, however, and no negative correction factor will be applied.

7.4 Final Design Basis Pond Temperature and Water Loss Computations-- Program UHS3

The final step in the cooling pond analysis is to combine the results of the programs COMET and UHS3 run from data provided by UHSPND, and the results of the manual statistical analyses to obtain a maximum water return temperature. Pond water loss is conservatively calculated manually.

Following the procedure of Section 6, two runs of UHS3 are made. The first run performs two simulations:

(1) Calculate the pond temperature in response only to the meteorologic variables with no emergency heat load.

(2) Calculate pond temperature in response only to heat load.

The input deck for the first run is shown in Figure 7.8. Notice that in the first INLIST input, TSKIP is set to a large time (5000 hours), to bypass the heat input table. The starting value of temperature TZERO is noncritical for this step and is set to 80°F. In the second INLIST input, TSKIP is reset to zero, and the values of K and E are chosen on the basis of experience to be 150 Btu/(ft² day)/°F and 90°F, respectively. Notice also that TZERO should be set equal to E which is 90°F.

The output from the first run is shown in Figure 7.9. If the mixed-tank model is chosen, the peak ambient pond temperature would be 86.32° F occurring 833 hours after the start. This is plotted graphically in Figure 7.10. The analysis would be similar if either the stratified- or plug-flow model had been chosen. The peak temperature due to pond heat load only would be 108.8°F or a rise of 18.8°F above the starting temperature of 90°F, and occurring at 191.6 hours after the start. This is plotted graphically in Figure 7.11.

50

PROGRAM TO COMPARE EQUILIBRIUM TEMPERATURES FROM TWO DATA SETS AND COMPUTE THE SENSITIVITY OF EACH VARIABLE

	DEW POINT (DEG. F)	DRY BULB	WIND SPEED (MPH)	SOLAR RAD. (BTU/FT**2/DY)	EQUILIBRIUM TEMP. (DEG. F)	EVAPORATION (FT**3/FT**2)
DATA SET 1	47.20	57.40	8.91	1381.60	64.80	.44
DATA SET 2	46.80	53.80	5.48	1381.60	66.99	.38

E2-E1 = 2.197   EVAP2-EVAP1 = -.06

DIFFERENCES IN E BETWEEN DATA SET 2 AND DATA SET 1 BY PARAMETER

DIFFERENCE DUE TO DEW POINT = -.155 DEG. F
DIFFERENCE DUE TO DRY BULB TEMP. = -1.200 DEG. F
DIFFERENCE DUE TO WIND SPEED = 3.481 DEG. F
DIFFERENCE DUE TO INSOLATION = .000 DEG. F
SUMMATION OF INDIVIDUAL DIFFERENCES = 2.126 DEG. F

****

	DEW POINT (DEG. F)	DRY BULB	WIND SPEED (MPH)	SOLAR RAD. (BTU/FT**2/DY)	EQUILIBRIUM TEMP. (DEG. F)	EVAPORATION (FT**3/FT**2)
DATA SET 1	62.70	72.80	6.12	1662.50	80.88	.59
DATA SET 2	61.30	67.50	3.90	1662.50	81.11	.53

E2-E1 = .228   EVAP2-EVAP1 = -.06

DIFFERENCES IN E BETWEEN DATA SET 2 AND DATA SET 1 BY PARAMETER

DIFFERENCE DUE TO DEW POINT = -.652 DEG. F
DIFFERENCE DUE TO DRY BULB TEMP. = -1.244 DEG. F
DIFFERENCE DUE TO WIND SPEED = 2.081 DEG. F
DIFFERENCE DUE TO INSOLATION = -.000 DEG. F
SUMMATION OF INDIVIDUAL DIFFERENCES = .185 DEG. F

Figure 7.6 Output of Program COMET.

```
*********************************************************
```

	DEW POINT (DEG. F)	DRY BULB	WIND SPEED (MPH)	SOLAR RAD. (BTU/FT**2/DY)	EQUILIBRIUM TEMP. (DEG. F)	EVAPORATION (FT**3/FT**2)
DATA SET 1	63.80	76.10	6.43	1888.20	83.54	.69
DATA SET 2	63.90	67.80	3.70	1888.20	84.58	.59

E2-E1 = 1.046   EVAP2-EVAP1 = -.10

DIFFERENCES IN E BETWEEN DATA SET 2 AND DATA SET 1 BY PARAMETER

```
DIFFERENCE DUE TO DEW POINT =            .046 DEG. F
DIFFERENCE DUE TO DRY BULB TEMP. =     -1.856 DEG. F
DIFFERENCE DUE TO WIND SPEED =          2.770 DEG. F
DIFFERENCE DUE TO INSOLATION =           .000 DEG. F
SUMMATION OF INDIVIDUAL DIFFERENCES =    .960 DEG. F
```

```
*********************************************************
```

	DEW POINT (DEG. F)	DRY BULB	WIND SPEED (MPH)	SOLAR RAD. (BTU/FT**2/DY)	EQUILIBRIUM TEMP. (DEG. F)	EVAPORATION (FT**3/FT**2)
DATA SET 1	64.60	75.40	5.90	1549.90	81.72	.56
DATA SET 2	64.80	69.40	3.20	1549.90	82.72	.49

E2-E1 = 1.001   EVAP2-EVAP1 = .07

DIFFERENCES IN E BETWEEN DATA SET 2 AND DATA SET 1 BY PARAMETER

```
DIFFERENCE DUE TO DEW POINT =            .099 DEG. F
DIFFERENCE DUE TO DRY BULB TEMP. =     -1.381 DEG. F
DIFFERENCE DUE TO WIND SPEED =          2.231 DEG. F
DIFFERENCE DUE TO INSOLATION =           .000 DEG. F
SUMMATION OF INDIVIDUAL DIFFERENCES =    .949 DEG. F
```

Figure 7.6 (Continued).

****

	DEW POINT (DEG. F)	DRY BULB	WIND SPEED (MPH)	SOLAR RAD. (BTU/FT**2/DY)	EQUILIBRIUM TEMP. (DEG. F)	EVAPORATION (FT**3/FT**2)
DATA SET 1	55.90	67.90	7.37	1309.30	72.45	.47
DATA SET 2	54.70	60.40	4.00	1309.30	72.75	.38

E2-E1 = .306   EVAP2-EVAP1 = -.09

DIFFERENCES IN E BETWEEN DATA SET 2 AND DATA SET 1 BY PARAMETER

DIFFERENCE DUE TO DEW POINT = -.530 DEG. F
DIFFERENCE DUE TO DRY BULB TEMP. = -2.123 DEG. F
DIFFERENCE DUE TO WIND SPEED = 2.863 DEG. F
DIFFERENCE DUE TO INSOLATION = .000 DEG. F
SUMMATION OF INDIVIDUAL DIFFERENCES = .211 DEG. F

****

	DEW POINT (DEG. F)	DRY BULB	WIND SPEED (MPH)	SOLAR RAD. (BTU/FT**2/DY)	EQUILIBRIUM TEMP. (DEG. F)	EVAPORATION (FT**3/FT**2)
DATA SET 1	46.70	63.50	8.61	1653.40	69.11	.59
DATA SET 2	45.00	56.70	5.60	1653.40	69.98	.49

E2-E1 = .867   EVAP2-EVAP1 = -.10

DIFFERENCES IN E BETWEEN DATA SET 2 AND DATA SET 1 BY PARAMETER

DIFFERENCE DUE TO DEW POINT = -.572 DEG. F
DIFFERENCE DUE TO DRY BULB TEMP. = -2.107 DEG. F
DIFFERENCE DUE TO WIND SPEED = 3.432 DEG. F
DIFFERENCE DUE TO INSOLATION = .000 DEG. F
SUMMATION OF INDIVIDUAL DIFFERENCES = .753 DEG. F

Figure 7.6 (Continued).

```
***********************************************************************

              DEW POINT   DRY BULB   WIND SPEED   SOLAR RAD.    EQUILIBRIUM TEMP.   EVAPORATION
              (DEG. F)                (MPH)        (BTU/FT**2/DY)    (DEG. F)        (FT**3/FT**2)

DATA SET 1      57.30       70.60      7.59         1687.00          76.58              .61

DATA SET 2      54.50       63.10      4.80         1687.00          76.47              .52

                                              E2-E1 = -.103      EVAP2-EVAP1 = -.09

DIFFERENCES IN E BETWEEN DATA SET 2 AND DATA SET 1 BY PARAMETER

DIFFERENCE DUE TO DEW POINT =               -1.162 DEG. F
DIFFERENCE DUE TO DRY BULB TEMP. =          -1.949 DEG. F
DIFFERENCE DUE TO WIND SPEED =               2.920 DEG. F
DIFFERENCE DUE TO INSOLATION =               -.000 DEG. F
SUMMATION OF INDIVIDUAL DIFFERENCES =        -.191 DEG. F

***********************************************************************

              DEW POINT   DRY BULB   WIND SPEED   SOLAR RAD.    EQUILIBRIUM TEMP.   EVAPORATION
              (DEG. F)                (MPH)        (BTU/FT**2/DY)    (DEG. F)        (FT**3/FT**2)

DATA SET 1      59.90       77.30      7.54         1766.70          79.99              .70

DATA SET 2      59.90       68.40      4.12         1766.70          81.45              .57

                                              E2-E1 = 1.463      EVAP2-EVAP1 = -.13

DIFFERENCES IN E BETWEEN DATA SET 2 AND DATA SET 1 BY PARAMETER

DIFFERENCE DUE TO DEW POINT =                -.000 DEG. F
DIFFERENCE DUE TO DRY BULB TEMP. =          -2.152 DEG. F
DIFFERENCE DUE TO WIND SPEED =               3.489 DEG. F
DIFFERENCE DUE TO INSOLATION =               -.000 DEG. F
SUMMATION OF INDIVIDUAL DIFFERENCES =        1.337 DEG. F
```

Figure 7.6 (Continued).

```
**************************************************

            DEW POINT   DRY BULB   WIND SPEED   SOLAR RAD.      EQUILIBRIUM TEMP.   EVAPORATION
            (DEG. F)               (MPH)        (BTU/FT**2/DY)  (DEG. F)            (FT**3/FT**2)

DATA SET 1    63.90      76.50       5.74         1386.50          80.45               .52

DATA SET 2    62.40      68.00       3.39         1386.50          79.53               .44

                                          E2-E1 = -.919      EVAP2-EVAP1 = -.08

DIFFERENCES IN E BETWEEN DATA SET 2 AND DATA SET 1 BY PARAMETER

DIFFERENCE DUE TO DEW POINT =         -.729 DEG. F
DIFFERENCE DUE TO DRY BULB TEMP. =  -2.020 DEG. F
DIFFERENCE DUE TO WIND SPEED =       1.786 DEG. F
DIFFERENCE DUE TO INSOLATION =       -.000 DEG. F
SUMMATION OF INDIVIDUAL DIFFERENCES = -.963 DEG. F

**************************************************

            DEW POINT   DRY BULB   WIND SPEED   SOLAR RAD.      EQUILIBRIUM TEMP.   EVAPORATION
            (DEG. F)               (MPH)        (BTU/FT**2/DY)  (DEG. F)            (FT**3/FT**2)

DATA SET 1    55.60      64.20       7.46         1199.70          70.25               .41

DATA SET 2    53.20      58.50       4.18         1199.70          70.22               .34

                                          E2-E1 = -.031      EVAP2-EVAP1 = -.06

DIFFERENCES IN E BETWEEN DATA SET 2 AND DATA SET 1 BY PARAMETER

DIFFERENCE DUE TO DEW POINT =         -1.085 DEG. F
DIFFERENCE DUE TO DRY BULB TEMP. =  -1.680 DEG. F
DIFFERENCE DUE TO WIND SPEED =       2.664 DEG. F
DIFFERENCE DUE TO INSOLATION =       -.000 DEG. F
SUMMATION OF INDIVIDUAL DIFFERENCES = -.100 DEG. F
```

Figure 7.6 (Continued).

```
*************************************************************************************************

           DEW POINT    DRY BULB    WIND SPEED    SOLAR RAD.      EQUILIBRIUM TEMP.    EVAPORATION
           (DEG. F)                 (MPH)         (BTU/FT**2/DY)  (DEG. F)             (FT**3/FT**2)

DATA SET 1   56.00       64.70        6.65         1563.70          74.47                 .51

DATA SET 2   53.60       61.00        4.36         1563.70          74.78                 .47


                                      E2-E1 =  .309      EVAP2-EVAP1 =  -.04
```

DIFFERENCES IN E BETWEEN DATA SET 2 AND DATA SET 1 BY PARAMETER

```
DIFFERENCE DUE TO DEW POINT        = -1.014 DEG. F
DIFFERENCE DUE TO DRY BULB TEMP.   =  -.998 DEG. F
DIFFERENCE DUE TO WIND SPEED       =  2.278 DEG. F
DIFFERENCE DUE TO INSOLATION       =   .000 DEG. F
SUMMATION OF INDIVIDUAL DIFFERENCES =  .266 DEG. F
```

```
*************************************************************************************************

           DEW POINT    DRY BULB    WIND SPEED    SOLAR RAD.      EQUILIBRIUM TEMP.    EVAPORATION
           (DEG. F)                 (MPH)         (BTU/FT**2/DY)  (DEG. F)             (FT**3/FT**2)

DATA SET 1   62.20       70.60        7.61         1636.60          78.42                 .57

DATA SET 2   59.90       65.70        4.53         1636.60          79.24                 .51


                                      E2-E1 =  .815      EVAP2-EVAP1 =  -.06
```

DIFFERENCES IN E BETWEEN DATA SET 2 AND DATA SET 1 BY PARAMETER

```
DIFFERENCE DUE TO DEW POINT        = -1.090 DEG. F
DIFFERENCE DUE TO DRY BULB TEMP.   = -1.204 DEG. F
DIFFERENCE DUE TO WIND SPEED       =  3.031 DEG. F
DIFFERENCE DUE TO INSOLATION       =   .000 DEG. F
SUMMATION OF INDIVIDUAL DIFFERENCES =  .737 DEG. F
```

Figure 7.6 (Continued).

```
************************************************************
```

	DEW POINT (DEG. F)	DRY BULB	WIND SPEED (MPH)	SOLAR RAD. (BTU/FT**2/DY)	EQUILIBRIUM TEMP. (DEG. F)	EVAPORATION (FT**3/FT**2)
DATA SET 1	66.30	75.00	6.84	1750.10	83.07	.63
DATA SET 2	63.90	69.40	3.54	1750.10	83.85	.56

E2-E1 = .780   EVAP2-EVAP1 = -.07

DIFFERENCES IN E BETWEEN DATA SET 2 AND DATA SET 1 BY PARAMETER

DIFFERENCE DUE TO DEW POINT =   -1.178 DEG. F
DIFFERENCE DUE TO DRY BULB TEMP. =   -1.248 DEG. F
DIFFERENCE DUE TO WIND SPEED =   3.123 DEG. F
DIFFERENCE DUE TO INSOLATION =   .000 DEG. F
SUMMATION OF INDIVIDUAL DIFFERENCES = .696 DEG. F

```
************************************************************
```

	DEW POINT (DEG. F)	DRY BULB	WIND SPEED (MPH)	SOLAR RAD. (BTU/FT**2/DY)	EQUILIBRIUM TEMP. (DEG. F)	EVAPORATION (FT**3/FT**2)
DATA SET 1	66.80	75.10	6.75	1517.80	81.75	.55
DATA SET 2	62.60	68.20	3.94	1517.80	80.52	.48

E2-E1 = -1.230   EVAP2-EVAP1 = -.07

DIFFERENCES IN E BETWEEN DATA SET 2 AND DATA SET 1 BY PARAMETER

DIFFERENCE DUE TO DEW POINT =   -2.110 DEG. F
DIFFERENCE DUE TO DRY BULB TEMP. =   -1.580 DEG. F
DIFFERENCE DUE TO WIND SPEED =   2.402 DEG. F
DIFFERENCE DUE TO INSOLATION =   .000 DEG. F
SUMMATION OF INDIVIDUAL DIFFERENCES = -1.288 DEG. F

Figure 7.6 (Continued).

```
********************************************************************************

              DEW POINT   DRY BULB   WIND SPEED   SOLAR RAD.      EQUILIBRIUM TEMP.   EVAPORATION
              (DEG. F)               (MPH)        (BTU/FT**2/DY)  (DEG. F)            (FT**3/FT**2)

DATA SET 1    56.30       62.80      7.31         1173.40         70.07               .38

DATA SET 2    52.90       57.90      4.47         1173.40         69.40               .33

                                                  E2-E1 = -.673   EVAP2-EVAP1 =  -.05


DIFFERENCES IN E BETWEEN DATA SET 2 AND DATA SET 1 BY PARAMETER

DIFFERENCE DUE TO DEW POINT =          -1.564 DEG. F
DIFFERENCE DUE TO DRY BULB TEMP. =     -1.444 DEG. F
DIFFERENCE DUE TO WIND SPEED =          2.288 DEG. F
DIFFERENCE DUE TO INSOLATION =          -.000 DEG. F
SUMMATION OF INDIVIDUAL DIFFERENCES =   -.721 DEG. F

********************************************************************************

SAMPLE R SQUARED FOR EQUILIBRIUM TEMP. =  .976        STANDARD ERROR =  .932 DEG.F
SAMPLE R SQUARED FOR EVAPORATION =  .956E+00          STANDARD ERROR =  .177E-01FT**3/FT**2
AVERAGE E, DATA SET 1 =  76.502
AVERAGE E, DATA SET 2 =  76.906
AVERAGE E2 - AVERAGE E1 =   .4036
AVERAGE EVAP2 - AVERAGE EVAP1 =  -.0762
```

Figure 7.6 (Continued).

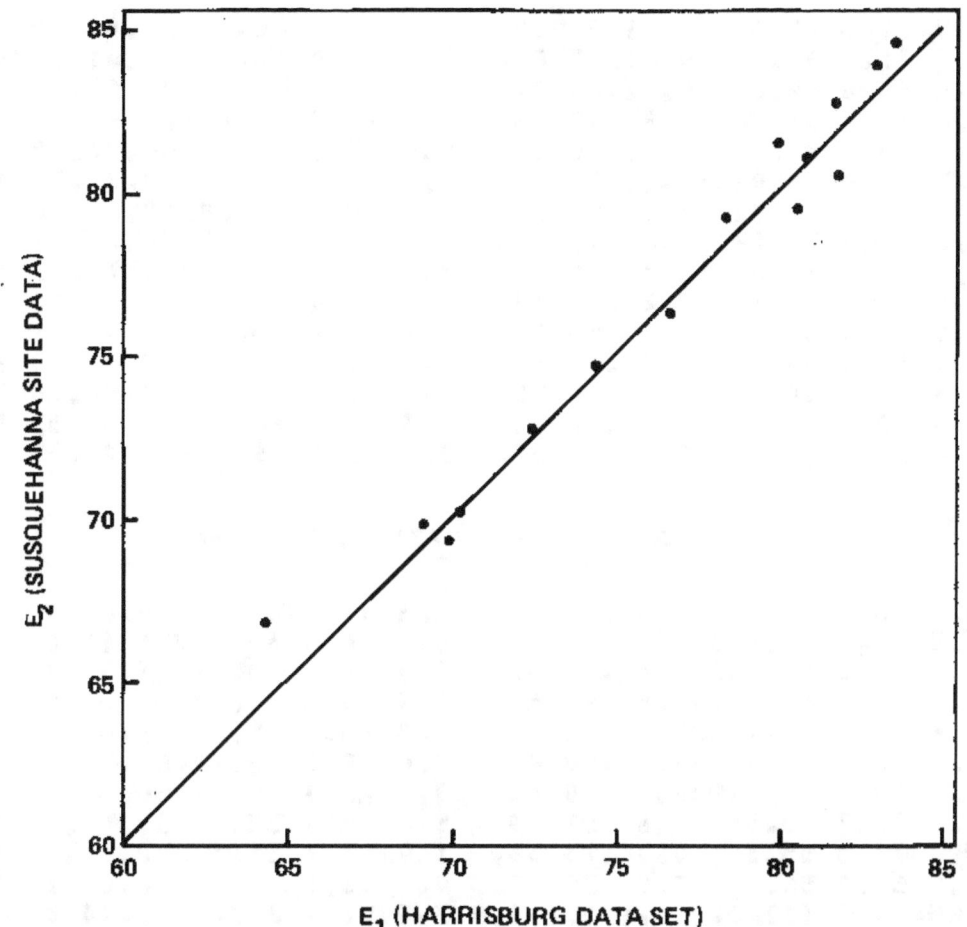

**Figure 7.7 Correlation of Equilibrium Temperatures, Susquehanna
River Site vs. Harrisburg Station.**

The second run of program UHS3 is set up after inspection of the first run.
The parameter TSKIP, which delays the start of the heat input table, is
adjusted so that the temperature peaks would be superimposed:

$$TSKIP = 833 - 191.6 = 641.4 \text{ hours}$$

The data deck for this run is shown in Figure 7.12. The output from this run
is shown in Figure 7.13, and shown graphically in Figure 7.14. The calculated
peak pond temperature is 105.22°F occurring at 810.6 hours after start. Note
that the predicted peak temperature by direct superposition of the preliminary
runs is 86.32°F + 18.8°F = 105.12°F, and would occur at 833.0 hours. The
close agreement is partially due to the good choice of K and E for the excess
temperature calculation in the first run, and must not be deemed to be
necessarily true in every case.

Because of nonlinearities in the heat transfer terms of the model, the true
maximum will not necessarily occur at the time predicted for direct
superposition. In fact, the calculated peak in the above example actually

```
960
 1  1.9 71.3 66.3    0.01.00 84.   0.0 71.0 66.0    0.01.00 84.
 2  1.5 71.0 66.0    0.01.00 84.   3.1 71.0 66.0    0.01.00 84.
 3  4.6 71.0 66.0    0.01.00 84.   5.0 71.3 66.7 134.61.00 85.
 4  5.4 71.7 67.3 406.21.00 86.   5.8 72.0 68.0 677.81.00 87.
 5  5.8 73.7 68.0 931.01.00 83.   5.8 75.3 68.01148.31.00 78.
 6  5.8 77.0 68.01315.11.00 74.   6.9 78.7 68.71759.8 .93 72.
 7  8.1 80.3 69.32128.5 .87 69.   9.2 82.0 70.02369.2 .80 67.
 8  8.8 82.3 70.02321.8 .77 66.   8.4 82.7 70.02134.0 .73 66.
 9  8.1 83.0 70.01812.7 .70 65.   8.1 82.0 69.71259.7 .73 66.
10  8.1 81.0 69.3 717.2 .77 68.   8.1 80.0 69.0 224.6 .80 69.
11  8.1 78.0 67.3    0.0 .63 70.   8.1 76.0 65.7    0.0 .47 70.
12  8.1 74.0 64.0    0.0 .30 71.   7.7 73.0 64.0    0.0 .20 74.
13  7.3 72.0 64.0    0.0 .10 76.   6.9 71.0 64.0    0.00.00 79.
14  4.6 70.3 64.3    0.0 .27 82.   2.3 69.7 64.7    0.0 .53 84.
15  0.0 69.0 65.0    0.0 .80 87.   3.1 70.3 65.7 213.4 .83 85.
```

*** CARDS 16 TO 470 ARE NOT SHOWN ***

```
471  3.5 66.0 61.0    0.01.00 84.   2.3 67.3 60.7  30.61.00 80.
472  1.2 68.7 60.3 300.91.00 75.   0.0 70.0 60.0 571.11.00 71.
473  0.0 71.3 60.01203.3 .87 68.   0.0 72.7 60.01931.3 .73 65.
474  0.0 74.0 60.02637.6 .60 62.   1.9 75.3 58.32960.5 .57 56.
475  3.8 76.7 56.73132.5 .53 51.   5.8 78.0 55.03133.4 .50 45.
476  7.7 78.7 55.32883.8 .50 45.   9.6 79.3 55.72486.7 .50 44.
477 11.5 80.0 56.01969.2 .50 44.   9.6 78.3 55.71160.4 .67 46.
478  7.7 76.7 55.3 471.6 .83 48.   5.8 75.0 55.0  30.61.00 50.
479  3.8 71.3 55.3    0.0 .73 58.   1.9 67.7 55.7    0.0 .47 67.
480  0.0 64.0 56.0    0.0 .20 75.   0.0 62.7 56.0    0.0 .13 79.
 $HFT NH=14,TH(1)=0,.01,1,1.1,1.9,3.9,5,8,12,24,29,140,840,2000,
    HEAT(1)=0,0,.85E9,2*.51E9,.5E9,.68E9,.6E9,.4E9,.31E9,.27E9,.21E9,
    .18E9,.1E9,FLOW(I)=14*3.6E5$
    RUN TO DETERMINE AMBIENT POND TEMPERATURE
 $INLIST VZERO=1.39392E7,A=1.7424E6,NSTEPS=4500,NPRINT=100,
    TZERO=80,TSKIP=5000,DT=0.2$
    RUN TO DETERMINE FORCED POND TEMPERATURE WITHOUT AMBIENT EFFECTS
 $INLIST TSKIP=0,IMET=1,AK1=150,E=90,TZERO=90,NPRINT=100$
    TERMINATE RUN
 $INLIST VZERO=0$
```

Figure 7.8 Data Deck for Program UHS3, First Set.

occurred about one day earlier than predicted (an error of a day is reasonable because of the variation of meteorology on a 1-day cycle). Table 7.1 illustrates the peak temperature predicted by varying the parameter TSKIP over a range of up to 30 hours.

The results indicate that, in this case, the maximum temperature is very nearly predicted at the time indicated by the direct superposition of the peaks.

VZERO	A	BLOW	AMAKE	
.13939E+08	.17424E+07	0.	0.	

NSTEPS	NPRINT	DT	TZERO	TSKIP
4500	100	.200	80.0	5000.0

QBASE	FBASE	E	AK1	IMET
0.	0.	80.0	150.0	0

BTA	BTD	BHS	BW
0.0	0.0	0.0	0.0

HEAT IN BTU/HR	TIME FROM START	FLOW IN FT**3/HR
0.	0.00	.360E+06
0.	.01	.360E+06
.850E+09	1.00	.360E+06
.510E+09	1.10	.360E+06
.510E+09	1.90	.360E+06
.500E+09	3.90	.360E+06
.680E+09	5.00	.360E+06
.600E+09	8.00	.360E+06
.400E+09	12.00	.360E+06
.310E+09	24.00	.360E+06
.270E+09	29.00	.360E+06
.210E+09	140.00	.360E+06
.180E+09	840.00	.360E+06
.100E+09	2000.00	.360E+06

Figure 7.9 Output From Program UHS3, First Set.

************** MODEL RESULTS **************

TIME		TEMPERATURE (F)			VOLUME	
HR	MIXED	STRAT	PLUG		FT**3	
20.0	79.9	79.9	79.9		.13920E+08	
40.0	78.7	79.6	78.7		.13895E+08	
60.0	77.0	78.5	77.0		.13868E+08	
80.0	73.8	77.1	73.8		.13828E+08	
100.0	68.2	74.8	68.2		.13765E+08	
120.0	66.1	70.0	66.1		.13737E+08	
140.0	65.7	68.2	65.7		.13723E+08	
160.0	65.7	67.3	65.7		.13713E+08	
180.0	65.9	67.1	65.9		.13705E+08	
200.0	66.3	67.2	66.3		.13692E+08	
220.0	68.3	67.8	68.3		.13682E+08	
240.0	70.2	69.6	70.2		.13672E+08	
260.0	70.3	70.6	70.3		.13661E+08	
280.0	70.7	70.3	70.8		.13650E+08	
300.0	70.8	70.6	70.9		.13635E+08	
320.0	72.1	70.8	72.2		.13622E+08	
340.0	73.2	72.4	73.2		.13612E+08	
360.0	74.5	73.2	74.5		.13597E+08	
380.0	74.5	74.3	74.5		.13569E+08	
400.0	72.5	73.7	72.5		.13541E+08	
420.0	70.8	72.1	70.8		.13519E+08	
440.0	70.1	70.8	70.1		.13503E+08	
460.0	70.7	70.6	70.7		.13488E+08	
480.0	71.3	71.3	71.3		.13472E+08	
500.0	71.6	71.5	71.6		.13458E+08	
520.0	73.1	71.3	73.1		.13444E+08	
540.0	74.0	72.4	74.0		.13431E+08	
560.0	75.1	73.3	75.1		.13420E+08	
580.0	75.9	75.0	75.9		.13409E+08	
600.0	75.8	75.6	75.8		.13383E+08	
620.0	77.4	75.5	77.4		.13371E+08	
640.0	78.9	76.7	78.9		.13356E+08	
660.0	79.4	77.5	79.4		.13340E+08	
680.0	79.0	78.2	79.0		.13322E+08	
700.0	79.6	78.6	79.6		.13308E+08	
720.0	80.6	79.1	80.6		.13299E+08	
740.0	82.5	80.0	82.5		.13288E+08	
760.0	83.7	81.5	83.7		.13273E+08	
780.0	83.8	82.4	83.9		.13259E+08	
800.0	83.6	82.8	83.6		.13240E+08	
820.0	84.6	83.3	84.6		.13220E+08	
840.0	85.2	84.4	85.2		.13193E+08	
860.0	84.8	84.6	84.8		.13159E+08	
880.0	83.8	84.0	83.8		.13125E+08	
900.0	80.0	82.8	80.0		.13068E+08	

MAXIMUM MODELLED TEMPERATURES:
MIXED MODEL =   85.90 AT   833.20 HOURS
STRAT MODEL =   84.69 AT   855.80 HOURS
PLUG MODEL =   85.91 AT   833.00 HOURS

Figure 7.9 (Continued).

62

RUN TO DETERMINE FORCED POND TEMPERATURE WITHOUT AMBIENT EFFECTS

VZERO	A	BLOW	AMAKE	
.13939E+08	.17424E+07	0.	0.	

NSTEPS	NPRINT	DT	TZERO	TSKIP
4500	100	.200	90.0	0.0

QBASE	FBASE	E	AK1	IMET
0.	0.	90.0	150.0	1

BTA	BTD	BHS	BW
0.0	0.0	0.0	0.0

HEAT IN BTU/HR	TIME FROM START	FLOW IN FT**3/HR
0.	0.00	.360E+06
0.	.01	.360E+06
.850E+09	1.00	.360E+06
.510E+09	1.10	.360E+06
.510E+09	1.90	.360E+06
.500E+09	3.90	.360E+06
.680E+09	5.00	.360E+06
.600E+09	8.00	.360E+06
.400E+09	12.00	.360E+06
.310E+09	24.00	.360E+06
.270E+09	29.00	.360E+06
.210E+09	140.00	.360E+06
.180E+09	840.00	.360E+06
.100E+09	2000.00	.360E+06

Figure 7.9 (Continued).

************* MODEL RESULTS *************

..TIME..	..TEMPERATURE (F)..			..VOLUME..
: HR :	MIXED :	STRAT :	PLUG :	FT**3 :
20.0 :	99.6 :	90.8 :	91.6 :	.13878E+08 :
40.0 :	103.2 :	99.4 :	100.4 :	.13799E+08 :
60.0 :	105.4 :	101.7 :	101.5 :	.13716E+08 :
80.0 :	106.8 :	103.9 :	103.5 :	.13620E+08 :
100.0 :	107.7 :	105.3 :	104.4 :	.13496E+08 :
120.0 :	108.3 :	106.4 :	105.1 :	.13365E+08 :
140.0 :	108.6 :	107.1 :	105.4 :	.13240E+08 :
160.0 :	108.7 :	107.6 :	105.6 :	.13120E+08 :
180.0 :	108.8 :	107.9 :	105.7 :	.13006E+08 :
200.0 :	108.8 :	108.1 :	105.8 :	.12888E+08 :
220.0 :	108.7 :	108.2 :	105.8 :	.12776E+08 :
240.0 :	108.7 :	108.3 :	105.7 :	.12675E+08 :
260.0 :	108.6 :	108.3 :	105.7 :	.12576E+08 :
280.0 :	108.5 :	108.4 :	105.6 :	.12476E+08 :
300.0 :	108.4 :	108.3 :	105.5 :	.12371E+08 :
320.0 :	108.3 :	108.3 :	105.4 :	.12275E+08 :
340.0 :	108.3 :	108.2 :	105.4 :	.12184E+08 :
360.0 :	108.2 :	108.2 :	105.3 :	.12094E+08 :
380.0 :	108.1 :	108.1 :	105.2 :	.11976E+08 :
400.0 :	108.0 :	108.0 :	105.1 :	.11857E+08 :
420.0 :	107.9 :	108.0 :	105.1 :	.11741E+08 :
440.0 :	107.8 :	107.9 :	105.0 :	.11625E+08 :
460.0 :	107.7 :	107.8 :	104.9 :	.11513E+08 :
480.0 :	107.6 :	107.7 :	104.8 :	.11404E+08 :
500.0 :	107.6 :	107.7 :	104.8 :	.11301E+08 :
520.0 :	107.5 :	107.6 :	104.7 :	.11207E+08 :
540.0 :	107.4 :	107.5 :	104.7 :	.11114E+08 :
560.0 :	107.4 :	107.4 :	104.6 :	.11025E+08 :
580.0 :	107.3 :	107.4 :	104.5 :	.10940E+08 :
600.0 :	107.2 :	107.3 :	104.5 :	.10851E+08 :
620.0 :	107.2 :	107.3 :	104.4 :	.10765E+08 :
640.0 :	107.1 :	107.2 :	104.4 :	.10680E+08 :
660.0 :	107.1 :	107.1 :	104.3 :	.10599E+08 :
680.0 :	107.0 :	107.1 :	104.3 :	.10517E+08 :
700.0 :	107.0 :	107.0 :	104.3 :	.10438E+08 :
720.0 :	106.9 :	107.0 :	104.2 :	.10365E+08 :
740.0 :	106.9 :	106.9 :	104.2 :	.10291E+08 :
760.0 :	106.8 :	106.9 :	104.1 :	.10221E+08 :
780.0 :	106.8 :	106.8 :	104.1 :	.10151E+08 :
800.0 :	106.7 :	106.8 :	104.1 :	.10078E+08 :
820.0 :	106.7 :	106.7 :	104.0 :	.10005E+08 :
840.0 :	106.6 :	106.7 :	104.0 :	.99302E+07 :
860.0 :	106.6 :	106.7 :	103.9 :	.98446E+07 :
880.0 :	106.5 :	106.6 :	103.9 :	.97571E+07 :
900.0 :	106.3 :	106.5 :	103.8 :	.96482E+07 :

```
MAXIMUM MODELLED TEMPERATURES:
MIXED MODEL =    108.78 AT    194.20 HOURS
STRAT MODEL =    108.35 AT    272.20 HOURS
PLUG  MODEL =    105.03 AT    209.60 HOURS
```

Figure 7.9 (Continued).

RUN TO DETERMINE FORCED POND TEMPERATURE WITHOUT AMBIENT EFFECTS

VZERO	A	BLOW	AMAKE
.13939E+08	.17424E+07	0.	0.

NSTEPS	NPRINT	DT	TZERO	TSKIP
4500	100	.200	90.0	0.0

QBASE	FBASE	E	AK1	IMET
0.	0.	90.0	150.0	1

BTA	BTD	BHS	BW
0.0	0.0	0.0	0.0

HEAT IN BTU/HR	TIME FROM START	FLOW IN FT**3/HR
0.	0.00	.360E+06
0.	.01	.360E+06
.850E+09	1.00	.360E+06
.510E+09	1.10	.360E+06
.510E+09	1.90	.360E+06
.500E+09	3.90	.360E+06
.680E+09	5.00	.360E+06
.600E+09	8.00	.360E+06
.400E+09	12.00	.360E+06
.310E+09	24.00	.360E+06
.270E+09	29.00	.360E+06
.210E+09	140.00	.360E+06
.180E+09	840.00	.360E+06
.100E+09	2000.00	.360E+06

Figure 7.9 (Continued).

65

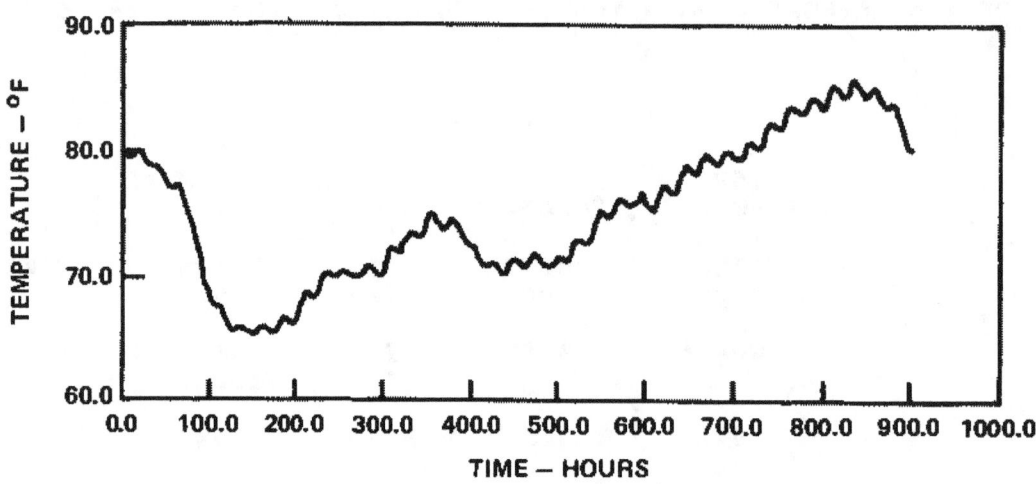

Figure 7.10 Ambient Pond Temperature as a Function of Time.

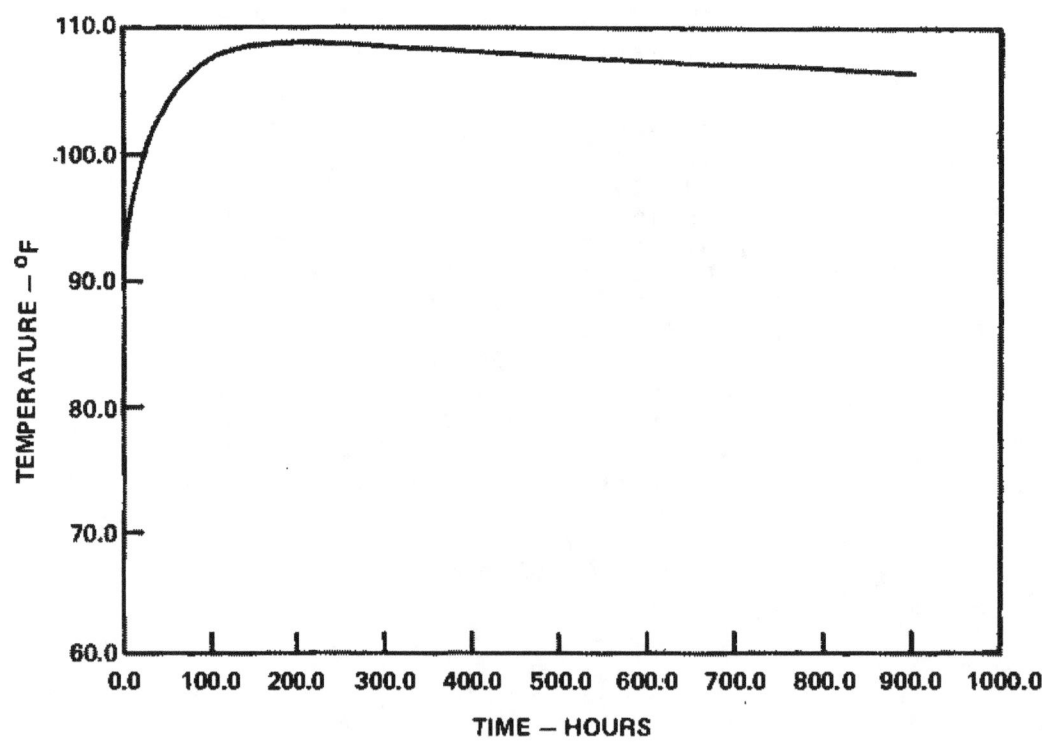

Figure 7.11 Pond Temperature With External Plant Heat Load and Constants E and K as a Function of Time.

```
960
1    1.9 71.3 66.3    0.01.00 84.    0.0 71.0 66.0    0.01.00 84.
2    1.5 71.0 66.0    0.01.00 84.    3.1 71.0 66.0    0.01.00 84.
3    4.6 71.0 66.0    0.01.00 84.    5.0 71.3 66.7 134.61.00 85.
4    5.4 71.7 67.3 406.21.00 86.    5.8 72.0 68.0 677.81.00 87.
5    5.8 73.7 68.0 931.01.00 83.    5.8 75.3 68.01148.31.00 78.
6    5.8 77.0 68.01315.11.00 74.    6.9 78.7 68.71759.8 .93 72.
7    8.1 80.3 69.32128.5 .87 69.    9.2 82.0 70.02369.2 .80 67.
8    8.8 82.3 70.02321.8 .77 66.    8.4 82.7 70.02134.0 .73 66.
9    8.1 83.0 70.01812.7 .70 65.    8.1 82.0 69.71259.7 .73 66.
10   8.1 81.0 69.3 717.2 .77 68.    8.1 80.0 69.0 224.6 .80 69.
11   8.1 78.0 67.3    0.0 .63 70.    8.1 76.0 65.7    0.0 .47 70.
12   8.1 74.0 64.0    0.0 .30 71.    7.7 73.0 64.0    0.0 .20 74.
13   7.3 72.0 64.0    0.0 .10 76.    6.9 71.0 64.0    0.00.00 79.
14   4.6 70.3 64.3    0.0 .27 82.    2.3 69.7 64.7    0.0 .53 84.
15   0.0 69.0 65.0    0.0 .80 87.    3.1 70.3 65.7 213.4 .83 85.
```

*** CARDS 16 TO 470 ARE NOT SHOWN ***

```
471  3.5 66.0 61.0    0.01.00 84.    2.3 67.3 60.7   30.61.00 80.
472  1.2 68.7 60.3 300.91.00 75.    0.0 70.0 60.0 571.11.00 71.
473  0.0 71.3 60.01203.3 .87 68.    0.0 72.7 60.01931.3 .73 65.
474  0.0 74.0 60.02637.6 .60 62.    1.9 75.3 58.32960.5 .57 56.
475  3.8 76.7 56.73132.5 .53 51.    5.8 78.0 55.03133.4 .50 45.
476  7.7 78.7 55.32883.8 .50 45.    9.6 79.3 55.72486.7 .50 44.
477 11.5 80.0 56.01969.2 .50 44.    9.6 78.3 55.71160.4 .67 46.
478  7.7 76.7 55.3 471.6 .83 48.    5.8 75.0 55.0   30.61.00 50.
479  3.8 71.3 55.3    0.0 .73 58.    1.9 67.7 55.7    0.0 .47 67.
480  0.0 64.0 56.0    0.0 .20 75.    0.0 62.7 56.0    0.0 .13 79.
$HFT NH=14,TH(1)=0,.01,1,1.1,1.9,3.9,5,8,12,24,29,140,840,2000,
    HEAT(1)=0,0,.85E9,2*.51E9,.5E9,.68E9,.6E9,.4E9,.31E9,.27E9,.21E9,
    .18E9,.1E9,FLOW(1)=14*3.6E5$
    RUN TO DETERMINE PEAK POND TEMPERATURE
$INLIST VZERO=1.39392E7,A=1.7424E6,NSTEPS=4500,NPRINT=100,TZERO=80,
    TSKIP=639.0,DT=0.2$
    TERMINATE RUN
$INLIST VZERO=0$
```

Figure 7.12 Data Deck for Program UHS3, Second Set.

7.4.1 Evaporative Loss

A conservative water loss calculation will be employed in which the maximum ambient 30-day water loss will be added to the 30-day seepage loss and the evaporative loss due to heat addition assuming 100% of the excess heat is lost by evaporation:

RUN TO DETERMINE PEAK POND TEMPERATURE

VZERO	A	BLOW	AMAKE	
.13939E+08	.17424E+07	0.	0.	

	NSTEPS	NPRINT	DT	TZERO	TSKIP
	4500	100	.200	80.0	639.0

	QBASE	FBASE	E	AK1	IMET
0.		0.	80.0	150.0	0

	BTA	BTD	BHS	BW
	0.0	0.0	0.0	0.0

HEAT IN BTU/HR	TIME FROM START	FLOW IN FT**3/HR
0.	0.00	.360E+06
0.	.01	.360E+06
.850E+09	1.00	.360E+06
.510E+09	1.10	.360E+06
.510E+09	1.90	.360E+06
.500E+09	3.90	.360E+06
.680E+09	5.00	.360E+06
.600E+09	8.00	.360E+06
.400E+09	12.00	.360E+06
.310E+09	24.00	.360E+06
.270E+09	29.00	.360E+06
.210E+09	140.00	.360E+06
.180E+09	840.00	.360E+06
.100E+09	2000.00	.360E+06

Figure 7.13 Output From Program UHS3, Second Set.

```
************ MODEL RESULTS ************
```

```
..TIME.......TEMPERATURE (F).........VOLUME....
:  HR   :  MIXED  :  STRAT  :  PLUG  :    FT**3    :
```

HR	MIXED	STRAT	PLUG	FT**3
20.0	79.9	79.9	79.9	.13920E+08
40.0	78.7	79.6	78.7	.13895E+08
60.0	77.0	78.5	77.0	.13868E+08
80.0	73.8	77.1	73.8	.13828E+08
100.0	68.2	74.8	68.2	.13765E+08
120.0	66.1	70.0	66.1	.13737E+08
140.0	65.7	68.2	65.7	.13723E+08
160.0	65.7	67.3	65.7	.13713E+08
180.0	65.9	67.1	65.9	.13705E+08
200.0	66.3	67.2	66.3	.13692E+08
220.0	68.3	67.8	68.3	.13682E+08
240.0	70.2	69.6	70.2	.13672E+08
260.0	70.3	70.6	70.3	.13661E+08
280.0	70.7	70.3	70.8	.13650E+08
300.0	70.8	70.6	70.9	.13635E+08
320.0	72.1	70.8	72.2	.13622E+08
340.0	73.2	72.4	73.2	.13612E+08
360.0	74.5	73.2	74.5	.13597E+08
380.0	74.5	74.3	74.5	.13569E+08
400.0	72.5	73.7	72.5	.13541E+08
420.0	70.8	72.1	70.8	.13519E+08
440.0	70.1	70.8	70.1	.13503E+08
460.0	70.7	70.6	70.7	.13488E+08
480.0	71.3	71.3	71.3	.13472E+08
500.0	71.6	71.5	71.6	.13458E+08
520.0	73.1	71.3	73.1	.13444E+08
540.0	74.0	72.4	74.0	.13431E+08
560.0	75.1	73.3	75.1	.13420E+08
580.0	75.9	75.0	75.9	.13409E+08
600.0	75.8	75.6	75.8	.13383E+08
620.0	77.4	75.5	77.4	.13371E+08
640.0	79.7	76.7	78.9	.13356E+08
660.0	89.8	78.6	81.8	.13325E+08
680.0	92.9	88.5	90.0	.13277E+08
700.0	95.9	90.8	91.9	.13230E+08
720.0	98.8	94.0	95.3	.13184E+08
740.0	101.9	96.6	98.6	.13133E+08
760.0	102.8	99.7	99.7	.13063E+08
780.0	103.3	100.5	100.1	.13003E+08
800.0	102.8	101.4	99.7	.12931E+08
820.0	103.8	101.8	100.7	.12863E+08
840.0	103.0	103.1	100.0	.12772E+08
860.0	102.1	102.2	99.1	.12685E+08
880.0	100.4	101.7	97.5	.12596E+08
900.0	94.9	99.8	92.0	.12477E+08

```
MAXIMUM MODELLED TEMPERATURES:
MIXED MODEL =   104.61 AT    810.80 HOURS
STRAT MODEL =   103.19 AT    836.20 HOURS
PLUG  MODEL =   100.77 AT    810.60 HOURS
```

Figure 7.13 (Continued).

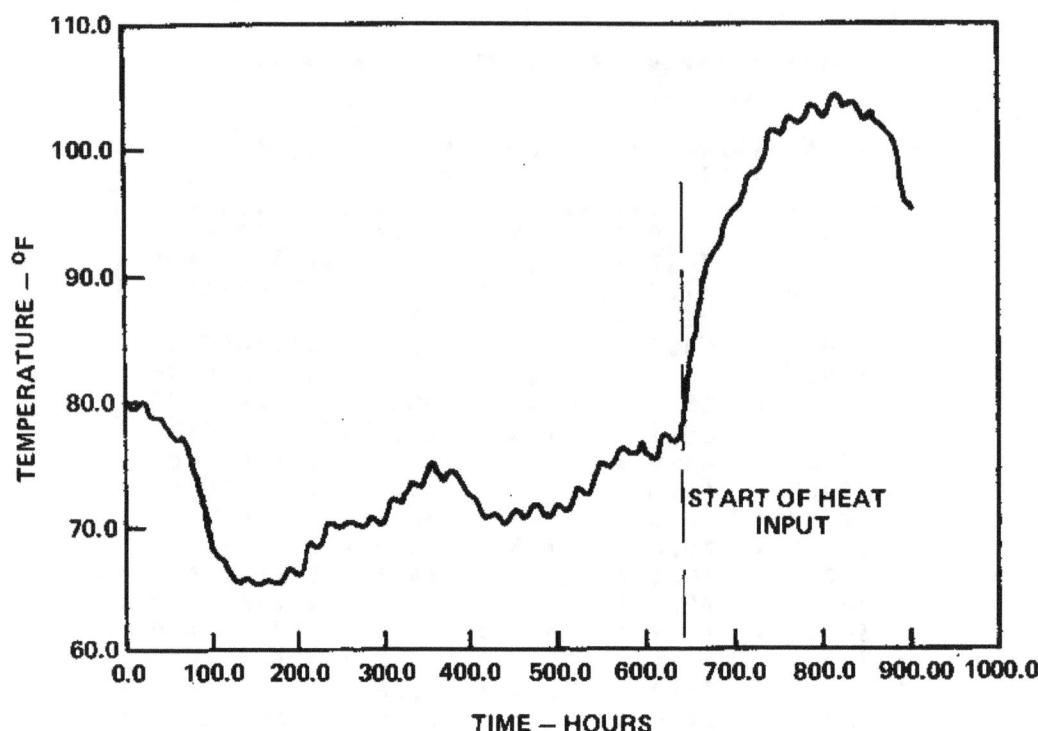

Figure 7.14 Pond Temperature (Final Calculation) as a Function of Time.

Maximum ambient 30-day loss $= 1.024 \times 10^6$ ft^3

$+$ 30-day seepage $= 1.44 \times 10^6$ ft^3

$+ \quad \dfrac{153 \times 10^9 \text{ Btu}}{1000 \text{ Btu/lb} \times 62.4 \text{ lb/ft}^3} \qquad = \underline{2.45 \times 10^6 \text{ ft}^3}$

Total 30-day water loss $\approx 4.91 \times 10^6$ ft^3

The total volume of the pond is 13.9×10^6 ft^3, so 65% of the pond water would be left after 30 days.

7.4.2 Correction Factors for Peak Temperature and Water Loss

To the peak temperature and 30-day water loss should be added the correction factors due to (1) statistical extrapolation of the offsite data to the 1% per year exceedence values and (2) the onsite versus offsite comparison of meteorological data.

The statistical extrapolation performed with the results of the program UHSPND indicate that the maximum ambient pond temperature and 30-day evaporation are greater than the extrapolated 1% per year exceedence values. Since only positive (conservative) correction factors are taken, no correction is necessary for (1).

70

Table 7.1 UHS3 Final Temperature Runs Varying
Starting Time

Run No.	TSKIP, hours	Hours from best estimate	Peak temperature, °F	Time of peak hours
1	641.4	0	105.22	810.6
2	617.4	-24	105.08	810.6
3	665.4	+24	105.29	810.8
4	629.4	-12	105.14	810.6
5	653.4	+12	105.26	810.8
6	635.4	- 6	105.19	810.6
7	647.4	+ 6	105.24	810.8
8	639	-2.4	105.21	810.6
9	643	+1.6	105.23	810.6

The onsite-offsite comparison of meteorological data from program COMET indicates that the onsite data would predict about a 0.4°F higher pond equilibrium temperature. Therefore, the maximum pond temperature should be raised accordingly:

Maximum pond temperature = 105.12°F + 0.4°F = 105.52°F.

Offsite evaporation is predicted to be higher than that onsite, so no correction factor for evaporation should be taken.

REFERENCES

1. U.S. Nuclear Regulatory Commission, Regulatory Guide 1.27, Revision 2, "Ultimate Heat Sinks for Nuclear Power Plants," January 1976.*

2. D. K. Brady, W. L. Graves, and J. C. Geyer, Surface Heat Exchange at Power Plant Cooling Lakes, Report No. 5, Edison Electric Institute, EEI Publication 69-401, New York, N.Y. 1969.

3. J. E. Edinger et al., "Generic Emergency Cooling Pond Analysis," May 1972, United States Atomic Energy Commission, Washington, D.C. October 1972.**

4. P. J. Ryan and D. R. F. Harleman, "An Analytical and Experimental Study of Transient Cooling Ponds Behavior," Report No. 161, R. M. Parsons Laboratory for Water Resources and Hydrodynamics, Department of Civil Engineering, Massachusetts Institute of Technology, Cambridge, Mass., January 1973.

5. G. H. Jirka, G. Abraham, and D. R. F. Harleman, "An Assessment of Techniques of Hydrothermal Prediction," Technical Report No. 203, R. M. Parsons Laboratory for Water Resources and Hydrodynamics, Department of Civil Engineering, Massachusetts Institute of Technology, Cambridge, Mass., July 1975 (also available as NUREG-0044**).

6. J. E. Edinger and E. M. Bushak, "A Hydrodynamic Two-Dimensional Resources Model," J. E. Edinger Associates, Wayne, Pa., 1975.

7. G. H. Jirka, D. W. Wood, and D. R. F. Harleman, "Transient Heat Releases From Offshore Nuclear Plants," Journal of the Hydraulics Division, American Society of Civil Engineers, No. HY2, 151-168, February 1977.

8. R. W.Hamon, L. L. Wiess, and W. T. Wilson, "Insolation as an Empirical Function of Daily Sunshine Duration," Monthly Weather Review 82(6):141-146, June 1954.

9. W. O. Wunderlich, "Heat and Mass Transfer Between a Water Surface and the Atmosphere," Report No. 14, TVA Engineering Laboratory, Norris, Tennessee, 1972.

10. L. R. Beard, "Statistical Methods in Hydrology," Engineering Memo. EM 1110-2-1450, U.S. Army Engineer District, Corps of Engineers, Sacramento, California, January 1962.

---

*Available for purchase from the NRC/GPO Sales Program, U.S. Nuclear Regulatory Commission, Washington, DC 20555.

**Available for purchase from the NRC/GPO Sales Program, U.S. Nuclear Regulatory Commission, Washington, DC 20555, and/or the National Technical Information Service, Springfield, VA 22161

APPENDIX A

Statistical Treatment of Output

Program UHSPND, in addition to determining the peak ambient pond temperature for the entire length of record, determines the maximum ambient temperature and evaporation for each year of the record. Subroutine SUB5 performs several simple manipulations of the yearly maximums to facilitate graphic analyses:

(1) The data are ranked from highest to lowest temperature.

(2) Their "probability" or plotting position is determined based on the number of years in the data set using the formulae (Ref. 10):

$$P_1 = 1 - (0.5)^{1/N} \qquad\qquad (A-1)$$

$$P_N = (0.5)^{1/N} \qquad\qquad (A-2)$$

$$P_i = P_1 - (i-1)\Delta P \qquad\qquad (A-3)$$

where

$$\Delta P = \frac{2(0.5)^{1/N}}{N-1}$$

where

N = number of data points in the set

P_1 = plotting position of the highest yearly maximum

P_N = plotting position of the lowest yearly maximum

P_i = plotting position of each individual point

(3) The first three moments of the distribution (mean, standard deviation, and skew) are determined from the formulae (Ref. 10):

$$M = \frac{\sum T}{N} \qquad\qquad \text{(mean)} \qquad\qquad (A-4)$$

$$s^2 = \frac{\sum T^2 - (\sum T)^2/N}{N-1} \qquad \text{(standard deviation)}^2 \qquad (A-5)$$

$$g = \frac{N^2 \sum T^3 - 3N\sum T\sum T^2 + 2(\sum T)^3}{N(N-1)(N-2)s^3} \qquad \text{(skew)} \qquad (A-6)$$

75

where

$\sum$ implies the sum over all N values in the data set

A.1 Additional Statistical Manipulation

The ranked annual maximum ambient temperature and evaporation data should be plotted in arithmetic-probability coordinates directly from the output from step 2 in order to get a qualitative look at the trends in the yearly maximum temperatures. A histogram of the same output may also be useful.

A maximum likelihood curve and error bands should be drawn on the probability graph following standard statistical procedures such as those outlined in Reference :

For convenience, several of the necessary tables and procedures described in this reference for Pearson type III coordinates are duplicated in the present report and will be described.

A.2 Maximum Likelihood Curve

The maximum likelihood frequency curve in probability coordinates is described by the following equation:

$$T = M + kS \qquad (A-7)$$

where

M = the mean

s = the standard deviation

k = a tabulated factor dependent on probability and skew

The procedures shown here involve only temperatures, but evaporation may be treated in exactly the same way. The maximum likelihood frequency curve should be computed using the following steps:

(1) Arbitrarily select values of P_∞, the probability of occurrence if the data set were drawn from an infinite population, to cover the range of interest on the graph. Suggested values would be P_∞ = 0.1, 1, 10, 50, 90, 99 and 99.9%.

(2) For each selected value of P_∞, find the k value corresponding to the adopted skew coefficient using Figure A.1. (Note: Beard suggests that skew cannot be reliably determined from small data sets, so zero skew is usually adopted.)

(3) Calculate T from Eq. (A-7) for each value of k determined.

(4) Find the corresponding value of P_N, the probability corrected for the limited size N of the data set, for each value of P_∞ using Figure A.2.

(5) Plot T vs P_N for each value selected on the same probability plot that the raw data were plotted.

76

PEARSON TYPE III COORDINATES

g (Skew coefficient)	k = Magnitude in standard deviations from mean for exceedence percentages of:												
	99.99	99.9	99	95	90	70	50	30	10	5	1.0	0.1	0.01
1.0	-1.88	-1.80	-1.59	-1.31	-1.12	-0.61	-0.16	0.38	1.34	1.87	3.03	4.54	5.92
0.8	-2.18	-2.03	-1.74	-1.38	-1.16	-0.60	-0.13	0.42	1.34	1.83	2.90	4.25	5.48
0.6	-2.53	-2.28	-1.88	-1.45	-1.19	-0.58	-0.09	0.45	1.33	1.79	2.77	3.96	5.04
0.4	-2.92	-2.54	-2.03	-1.51	-1.22	-0.57	-0.06	0.48	1.32	1.74	2.62	3.67	4.60
0.2	-3.32	-2.81	-2.18	-1.58	-1.25	-0.55	-0.03	0.51	1.30	1.69	2.48	3.38	4.16
0.0	-3.73	-3.09	-2.33	-1.64	-1.28	-0.52	0.00	0.52	1.28	1.64	2.33	3.09	3.73
-0.2	-4.16	-3.38	-2.48	-1.69	-1.30	-0.51	0.03	0.55	1.25	1.58	2.18	2.81	3.32
-0.4	-4.60	-3.67	-2.62	-1.74	-1.32	-0.48	0.06	0.57	1.22	1.51	2.03	2.54	2.92
-0.6	-5.04	-3.96	-2.77	-1.79	-1.33	-0.45	0.09	0.58	1.19	1.45	1.88	2.28	2.53
-0.8	-5.48	-4.25	-2.90	-1.83	-1.34	-0.42	0.13	0.60	1.16	1.38	1.74	2.03	2.18
-1.0	-5.92	-4.54	-3.03	-1.87	-1.34	-0.38	0.16	0.61	1.12	1.31	1.59	1.80	1.88
Skew Coefficients Commonly Used													
.00	-3.73	-3.09	-2.33	-1.64	-1.28	-0.52	0.00	0.52	1.28	1.64	2.33	3.09	3.73
-.04	-3.82	-3.15	-2.36	-1.65	-1.28	-0.52	0.01	0.53	1.27	1.63	2.30	3.03	3.65
-.12	-3.99	-3.26	-2.42	-1.67	-1.29	-0.51	0.02	0.54	1.26	1.60	2.24	2.92	3.48
-.23	-4.23	-3.42	-2.50	-1.70	-1.30	-0.50	0.03	0.55	1.25	1.57	2.16	2.77	3.26
-.32	-4.42	-3.55	-2.56	-1.72	-1.31	-0.49	0.05	0.56	1.23	1.54	2.09	2.68	3.08
-.37	-4.53	-3.63	-2.60	-1.73	-1.32	-0.48	0.06	0.57	1.22	1.52	2.05	2.58	2.98
-.40	-4.60	-3.67	-2.62	-1.74	-1.32	-0.48	0.06	0.57	1.22	1.51	2.03	2.54	2.92

NOTE: Approximate transformations between normal deviate (X) and Pearson Type III deviate k can be accomplished with the following equation:

$$k = \frac{2}{g}\left\{\left[\frac{g}{6}\left(x - \frac{g}{6}\right) + 1\right]^3 - 1\right\}$$

Figure A.1 Pearson Type III Coordinates (After Ref. 10, Exhibit 39).

TABLE OF P_N VERSUS P_∞ IN PERCENT

For use with samples drawn from a normal population

P_∞ / N-1	50.0	30.0	10.0	5.0	1.0	0.1	0.01
1	50.0	37.2	24.3	20.4	15.4	12.1	10.2
2	50.0	34.7	19.3	14.6	9.0	5.7	4.3
3	50.0	33.6	16.9	11.9	6.4	3.5	2.3
4	50.0	33.0	15.4	10.4	5.0	2.4	1.37
5	50.0	32.5	14.6	9.4	4.2	1.79	.92
6	50.0	32.2	13.8	8.8	3.6	1.38	.66
7	50.0	31.9	13.5	8.3	3.2	1.13	.50
8	50.0	31.7	13.1	7.9	2.9	.94	.39
9	50.0	31.6	12.7	7.6	2.7	.82	.31
10	50.0	31.5	12.5	7.3	2.5	.72	.25
11	50.0	31.4	12.3	7.1	2.3	.64	.21
12	50.0	31.3	12.1	6.9	2.2	.58	.18
13	50.0	31.2	11.9	6.8	2.1	.52	.16
14	50.0	31.1	11.8	6.7	2.0	.48	.14
15	50.0	31.1	11.7	6.6	1.96	.45	.13
16	50.0	31.0	11.6	6.5	1.90	.42	.12
17	50.0	31.0	11.5	6.4	1.84	.40	.11
18	50.0	30.9	11.4	6.3	1.79	.38	.10
19	50.0	30.9	11.3	6.2	1.74	.36	.091
20	50.0	30.8	11.3	6.2	1.70	.34	.084
21	50.0	30.8	11.2	6.1	1.67	.33	.078
22	50.0	30.8	11.1	6.1	1.63	.31	.073
23	50.0	30.7	11.1	6.0	1.61	.30	.068
24	50.0	30.7	11.0	6.0	1.58	.29	.064
25	50.0	30.7	11.0	5.9	1.55	.28	.060
26	50.0	30.6	10.9	5.9	1.53	.27	.057
27	50.0	30.6	10.9	5.9	1.51	.26	.054
28	50.0	30.6	10.9	5.8	1.49	.26	.051
29	50.0	30.6	10.8	5.8	1.47	.25	.049
30	50.0	30.6	10.8	5.8	1.45	.24	.046
40	50.0	30.4	10.6	5.6	1.33	.20	.034
60	50.0	30.3	10.4	5.4	1.22	.16	.025
120	50.0	30.2	10.2	5.2	1.11	.13	.017
∞	50.0	30.0	10.0	5.0	1.00	.10	.010

NOTE: P_N values above are usable approximately with Pearson Type III distributions having small skew coefficients.

Figure A. 2 Table of P_N Versus P_∞ in Percent (After Ref. 10, Exhibit 40).

A.3   Error Bands

Error bands for the 5% and the 95% confidence limits may also be plotted using the following procedure:

(1)   For the same values of P_∞ selected in the computation of the maximum likelihood frequency curve, select the error of estimation from Figure A.3 for the 0.05 and 0.95 levels of confidence ε_5 and ε_{95}, respectively.

(2)   Determine the coordinate of the error band lines using the formulae:

$$T_{0.95} = M + (k+\varepsilon_{95})s \qquad\qquad (A\text{-}8)$$

$$T_{0.05} = M + (k+\varepsilon_5)s \qquad\qquad (A\text{-}9)$$

for each value of P_∞.

(3)   Plot $T_{0.95}$ and $T_{0.05}$ vs P_∞ on the same plot as the maximum likelihood curve and the raw data.   (Note: Do <u>not</u> plot $T_{0.95}$ and $T_{0.05}$ vs P_N as in the maximum likelihood curve.)

The error limit curves express the probability of a value falling outside of the error bands in any given year.   For the 95% and 5% bands, therefore, there is 1 chance in 20 that the ambient temperature value for any given recurrence interval is greater than indicated by the 5% curve and 1 chance in 20 that it is less than the 95% curve.

An example of the statistical procedure is offered in Section 7.

The maximum likelihood curves for temperature T and 30-day evaporation rate W_e are extrapolated to the 100-year recurrence interval (0.01 probability per year) to determine T_{100} and W_{100}.*   Correction factors for peak temperature ΔT and evaporation ΔWe are determined by comparing T_{100} and W_{100} with their corresponding highest observed values from the record, T_{max} and W_{max}:

$$\Delta T = T_{100} - T_{max} \qquad\qquad (A\text{-}10)$$

$$\Delta W_e = W_{100} - W_{max} \qquad\qquad (A\text{-}11)$$

Only correction factors greater than zero are considered.   If the maximum observed temperature or evaporation is higher than the 100-year recurrence values, no correction factor is taken.   These correction factors may be added directly to the peak loaded pond temperature and evaporations determined in subsequent calculations.

---

*Other recurrence intervals may be used.

ERRORS OF ESTIMATED VALUES

As Coefficients of Standard Deviation

Level of Significance*	Years of Record (N)	Exceedence Frequency in Percent						
		0.1	1	10	50	90	99	99.9
.05	5	4.41	3.41	2.12	.95	.76	1.00	1.22
	10	2.11	1.65	1.07	.58	.57	.76	.94
	15	1.52	1.19	.79	.46	.48	.65	.80
	20	1.23	.97	.64	.39	.42	.58	.71
	30	.93	.74	.50	.31	.35	.49	.60
	40	.77	.61	.42	.27	.31	.43	.53
	50	.67	.54	.36	.24	.28	.39	.49
	70	.55	.44	.30	.20	.24	.34	.42
	100	.45	.36	.25	.17	.21	.29	.37
.25	5	1.41	1.09	.68	.33	.31	.41	.49
	10	.77	.60	.39	.22	.24	.32	.39
	15	.57	.45	.29	.18	.20	.27	.34
	20	.47	.37	.25	.15	.18	.24	.30
	30	.36	.29	.19	.12	.15	.20	.25
	40	.30	.24	.16	.11	.13	.18	.22
	50	.27	.21	.14	.10	.12	.16	.20
	70	.22	.17	.12	.08	.10	.14	.18
	100	.18	.14	.10	.07	.09	.12	.15
.75	5	- .49	- .41	- .31	-.33	- .68	-1.09	-1.41
	10	- .39	- .32	- .24	-.22	- .39	- .60	- .77
	15	- .34	- .27	- .20	-.18	- .29	- .45	- .57
	20	- .30	- .24	- .18	-.15	- .25	- .37	- .47
	30	- .25	- .20	- .15	-.12	- .19	- .29	- .36
	40	- .22	- .18	- .13	-.11	- .16	- .24	- .30
	50	- .20	- .16	- .12	-.10	- .14	- .21	- .27
	70	- .18	- .14	- .10	-.08	- .12	- .17	- .22
	100	- .15	- .12	- .09	-.07	- .10	- .14	- .18
.95	5	-1.22	-1.00	- .76	-.95	-2.12	-3.41	-4.41
	10	- .94	- .76	- .57	-.58	-1.07	-1.65	-2.11
	15	- .80	- .65	- .48	-.46	- .79	-1.19	-1.52
	20	- .71	- .58	- .42	-.39	- .64	- .97	-1.23
	30	- .60	- .49	- .35	-.31	- .50	- .74	- .93
	40	- .53	- .43	- .31	-.27	- .42	- .61	- .77
	50	- .49	- .39	- .28	-.24	- .36	- .54	- .67
	70	- .42	- .34	- .24	-.20	- .30	- .44	- .55
	100	- .37	- .29	- .21	-.17	- .25	- .36	- .45

* Chance of true value being greater than sum of normal-curve value and given error.

Figure A.3 Errors of Estimated Values (After Ref. 10, Exhibit 6).

APPENDIX B

Computer Codes

```
      PROGRAM UHSPND(INPUT,OUTPUT,TAPE9,TAPE8=/495,TAPE5=INPUT        ULTSINK2
     1,TAPE6=OUTPUT,PUNCH,TAPE4=PUNCH)                                ULTSINK3
C                                                                     ULTSINK4
C     PROGRAM UHSPND IS A PROGRAM UNDER DEVELOPMENT BY THE STAFF OF THE ULTSINK5
C     HYDROLOGIC ENGINEERING SECTION OF THE U.S. NUCLEAR REGULATORY    ULTSINK6
C     COMMISSION FOR USE IN EVALUATING THE DESIGN BASIS METEOROLOGY OF ULTSINK7
C     SMALL COOLING PONDS USED AS THE ULTIMATE HEAT SINK OF A NUCLEAR  ULTSINK8
C     POWER PLANT.  THE PROGRAM USES HISTORICAL WEATHER DATA PROVIDED  ULTSINK9
C     ON TAPE BY THE NATIONAL WEATHER SERVICE AND A SIMPLIFIED POND    ULTSIN10
C     TEMPERATURE MODEL TO DETERMINE THE PERIOD OF RECORD WHICH WOULD  ULTSIN11
C     RESULT IN EITHER THE  LOWEST COOLING PERFORMANCE OR HIGHEST      ULTSIN12
C     EVAPORATIVE WATER LOSS IN A GIVEN POND.  THE USE OF THE PROGRAM  ULTSIN13
C     AND THE ANALYTICAL TECHNIQUES WHICH IT EMPLOYS ARE FULLY DESCRIBEDULTSIN14
C     IN LITERATURE AVAILABLE THROUGH THE HYDROLOGIC ENGINEERING       ULTSIN15
C     SECTION.  ALL QUESTIONS AND COMMENTS SHOULD BE ADDRESSED TO      ULTSIN16
C     R. CODELL.                                                       ULTSIN17
C                                                                      ULTSIN18
      REAL LAT1,LAT,YRMODY(3),YRMAX(40,8)                              JULY9   1
      LAT1=0.                                                          ULTSIN21
      WRITE(6,100)                                                     ULTSIN22
  100 FORMAT(1H1,20(/),10X,'U.S. NUCLEAR REGULATORY COMMISSION- ULTIMATEULTSIN23
     1 HEAT SINK COOLING POND METEOROLOGICAL SCANNING MODEL',/10X,'R CODULTSIN24
     2ELL AND W NUTTLE, NOVEMBER 1979',/1H1)                           ULTSIN25
      NAMELIST/INPUT/N,A,V,LAT,ISRCH,IPRNT,YRMODY                      ULTSIN26
      DATA N,ISRCH,IPRNT/1,1,0/                                        ULTSIN27
C                                                                      ULTSIN28
C     READ DATA CARD                                                   ULTSIN29
C                                                                      ULTSIN30
    1 READ(5,INPUT)                                                    ULTSIN31
      IF(N.EQ.0) STOP                                                  ULTSIN32
C                                                                      ULTSIN33
C     IF THIS IS THE FIRST DATA CARD OR IF LAT HAS CHANGED, GENERATE A ULTSIN34
C     NEW INTERMEDIATE FILE.                                           ULTSIN35
C                                                                      ULTSIN36
      IF(ABS(LAT1-LAT).GE..001) CALL SUB1(LAT)                         ULTSIN37
      LAT1=LAT                                                         ULTSIN38
      IF(N.GT.99) GO TO 4                                              ULTSIN39
      IF(V.LT.0.)V=V*(-43560.)                                         ULTSIN40
      IF(A.LT.0.)A=A*(-43560.)                                         ULTSIN41
      A1=A/43560.                                                      ULTSIN42
      V1=V/43560.                                                      ULTSIN43
C                                                                      ULTSIN44
C     PRINT POND PARAMETERS.                                           ULTSIN45
C                                                                      ULTSIN46
      WRITE(6,510)N,A,A1,V,V1,ISRCH,IPRNT                              ULTSIN47
  510 FORMAT(5(/),T20,10('*'),' POND NUMBER ',I2,' HAS THE FOLLOWING PARULTSIN48
     1AMETERS ',25('*'),//,T35,'SURFACE AREA'2X,F12.2,' FT**2 (',F9.2,  ULTSIN49
     2' ACRES)',//,T35,'VOLUME',8X,F12.2,' FT**3 (',F9.2,' ACRE-FT)',//,ULTSIN50
     3T35,'ISRCH = ',I2,T65,'IPRNT = ',I2)                             ULTSIN51
      WRITE(6,550)N                                                    ULTSIN52
  550 FORMAT(5(/),T20,10('*'),' POND NUMBER ',I2,' HAS BEEN MODELLED TO ULTSIN53
     1DETERMINE THE WORST ',13('*'),/,T38,  'PERIODS FOR COOLING AND EVAULTSIN54
     2PORATIVE WATER LOSS',/,1H1)                                      ULTSIN55
C                                                                      ULTSIN56
C     MODEL TO FIND YEARLY MAXIMUM TEMPERATURES AND 30 DAY EVAPORATIVE ULTSIN57
C     LOSSES.                                                          ULTSIN58
C                                                                      ULTSIN59
      CALL SUB2(A,V,YRMAX)                                             JULY9   2
C                                                                      ULTSIN61
C     RANK YEARLY MAXIMUM TEMPERATURES AND 30 DAY EVAPORATIVE LOSSES\  ULTSIN62
C     COMPUTE 100 YEAR EXCEEDENCES, SAMPLE MEANS, STANDARD DEVIATIONS, ULTSIN63
C     AND SKEWS.                                                       ULTSIN64
```

Figure B.1 Listing of Program UHSPND.

```
C                                                                    ULTSIN65
      CALL SUB5(YRMAX)                                               ULTSIN66
      IF(ISRCH.LE.0.OR.ISRCH.GE.6) GO TO 1                           ULTSIN67
C                                                                    ULTSIN68
C     PRINT AND/OR PUNCH DAILY METEOROLOGY FOR THE PERIODS OF RECORD ULTSIN69
C     PRECEEDING THE HIGHEST ISRCH POND TEMPERATURES. (ISRCH ) 6)    ULTSIN70
C                                                                    ULTSIN71
      DO 2 I=1,ISRCH                                                 ULTSIN72
      DO 3 J=1,3                                                     ULTSIN73
      J1=J+1                                                         ULTSIN74
    3 YRMODY(J)=YRMAX(I,J1)                                          ULTSIN75
      CALL SUB3(YRMODY,IPRNT)                                        JULY9  3
      IF(IPRNT.EQ.1) WRITE(6,520)                                    ULTSIN77
  520 FORMAT(1H1)                                                    ULTSIN78
    2 CONTINUE                                                       ULTSIN92
      GO TO 1                                                        ULTSIN93
    4 YRMODY(3)=1.                                                   ULTSIN94
C                                                                    ULTSIN95
C     CALCULATE AND PRINT MONTHLY AVERAGES OF EACH PARAMETER IN METABL. ULTSIN96
C                                                                    ULTSIN97
      CALL SUB4(YRMODY,LAT )                                         ULTSIN98
      GO TO 1                                                        ULTSIN99
      END                                                            ULTSI100
      SUBROUTINE SUB1(LAT)                                           SUB1   2
C                                                                    SUB1   3
C                                                                    SUB1   4
      REAL METABL(27,10),SRAD(25),LAT                                SUB1   5
      COMMON IDATE(3), IHOUR(6),WINDSP(6),TEMPDB(6),TEMPWB(6),TEMPDP(6),SUB1 6
     1HUMID(6),PRESSR(6),SKY(6)                                      SUB1   7
      DATA METABL/270*0./                                            SUB1   8
      DATA SRAD /25*0./                                              SUB1   9
      WRITE(6,520) LAT                                               SUB1  10
  520 FORMAT(5(/),T20,10('*'),' SUBROUTINE SUB1 HAS BEEN CALLED FOR LATISUB1 11
     1TUDE = ',F5.2,' DEG. NORTH ',5('*'),/)                         SUB1  12
C                                                                    SUB1  13
C     POSITION TAPE TO FIRST OF MAY.                                 SUB1  14
C                                                                    SUB1  15
      CALL READRC                                                    SUB1  16
      I=(121-IDATE(3))*4-2                                           SUB1  17
      DO 2 J=1,I                                                     SUB1  18
    2 READ(8)                                                        SUB1  19
    3 CALL READRC                                                    SUB1  20
      IF(IHOUR(1).NE.0) GO TO 3                                      SUB1  21
      IF(IDATE(2).LT.5) GO TO 3                                      SUB1  22
C                                                                    SUB1  23
C     READ IN FIRST 6 LINES OF DATA                                  SUB1  24
C                                                                    SUB1  25
      DO 4 I=1,6                                                     SUB1  26
      METABL(I,1)=IDATE(1)                                           SUB1  27
      METABL(I,2)=IDATE(2)                                           SUB1  28
      METABL(I,3)=IDATE(3)                                           SUB1  29
      METABL(I,4)=IHOUR(I)                                           SUB1  30
      METABL(I,5)=WINDSP(I)                                          SUB1  31
      METABL(I,6)=TEMPDB(I)                                          SUB1  32
      METABL(I,7)=TEMPDP(I)                                          SUB1  33
      METABL(I,8)=SKY(I)                                             SUB1  34
      METABL(I,9)=SKY(I)                                             SUB1  35
    4 METABL(I,10)=HUMID(I)                                          SUB1  36
C                                                                    SUB1  37
C     MAKE SURE THAT THE FIRST LINE OF DATA IS COMPLETE.             SUB1  38
C     IF DATA ARE MISSING, SUBSTITUTE FROM THE SECOND OR THIRD LINES SUB1  39
C     IF FIRST THREE LINES ARE BAD, SKIP TO THE NEXT DAY.            SUB1  40
C                                                                    SUB1  41
```

Figure B.1 (Continued).

```
      INDEX=1                                              SUB1   42
      IYR=IDATE(1)                                         SUB1   43
      IMON=IDATE(2)                                        SUB1   44
      IDAY=IDATE(3)                                        SUB1   45
      I=1                                                  SUB1   46
      GO TO 6                                              SUB1   47
    5 IF(I.EQ.3) GO TO 12                                  SUB1   48
      I=I+1                                                SUB1   49
      DO 7 J=5,10                                          SUB1   50
    7 IF(METABL(1,J).GE.999.) METABL(1,J)=METABL(I,J)      SUB1   51
    6 DO 1 J=5,10                                          SUB1   52
      IF(METABL(1,J).GE.9999.) GO TO 5                     SUB1   53
    1 CONTINUE                                             SUB1   54
      INDEX=2                                              SUB1   55
C                                                          SUB1   56
C     READ IN REST OF FIRST DAY'S DATA.                    SUB1   57
C                                                          SUB1   58
      DO 8 K=7,19,6                                        SUB1   59
      K5=K+5                                               SUB1   60
      CALL READRC                                          SUB1   61
      DO 8 J=K,K5                                          SUB1   62
      IK1=I-K+1                                            SUB1   63
      DO 8 I=K,K5                                          SUB1   64
      IK1=I-K+1                                            SUB1   65
      METABL(I,1)=IDATE(1)                                 SUB1   66
      METABL(I,2)=IDATE(2)                                 SUB1   67
      METABL(I,3)=IDATE(3)                                 SUB1   68
      METABL(I,4)=IHOUR(IK1)                               SUB1   69
      METABL(I,5)=WINDSP(IK1)                              SUB1   70
      METABL(I,6)=TEMPDB(IK1)                              SUB1   71
      METABL(I,7)=TEMPDP(IK1)                              SUB1   72
      METABL(I,8)=SKY(IK1)                                 SUB1   73
      METABL(I,9)=SKY(IK1)                                 SUB1   74
    8 METABL(I,10)=HUMID(IK1)                              SUB1   75
      CALL READRC                                          SUB1   76
      DO 9 I=1,3                                           SUB1   77
      I24=I+24                                             SUB1   78
      METABL(I24,1)=IDATE(1)                               SUB1   79
      METABL(I24,2)=IDATE(2)                               SUB1   80
      METABL(I24,3)=IDATE(3)                               SUB1   81
      METABL(I24,4)=IHOUR(I)                               SUB1   82
      METABL(I24,5)=WINDSP(I)                              SUB1   83
      METABL(I24,6)= TEMPDB(I)                             SUB1   84
      METABL(I24,7)=TEMPDP(I)                              SUB1   85
      METABL(I24,8)=SKY(I)                                 SUB1   86
      METABL(I24,9)=SKY(I)                                 SUB1   87
    9 METABL(I24,10)=HUMID(I)                              SUB1   88
      METABL(25,4)=24.                                     SUB1   89
C                                                          SUB1   90
C     SEARCH DATA RECORD FOR MISSING DATA AND INTERPOLATE TO SUB1 91
C     COMPLETE RECORD.                                     SUB1   92
C                                                          SUB1   93
      DO 10 I=1,25                                         SUB1   94
      DO 10 K=5,10                                         SUB1   95
      IF (METABL(I,K).LT.9999.) GO TO 10                   SUB1   96
      I1=I+1                                               SUB1   97
      IF(METABL(I1,K).GE.9999.) GO TO 11                   SUB1   98
      I0=I-1                                               SUB1   99
      METABL(I,K)=METABL(I1,K)-(METABL(I1,K)-METABL(I0,K))*.5 SUB1 100
      GO TO 10                                             SUB1  101
   11 I2=I+2                                               SUB1  102
C                                                          SUB1  103
C     IF THREE OR MORE CONSECUTIVE HOURS OF DATA ARE MISSING, SKIP SUB1 104
```

Figure B.1 (Continued).

85

```
C        TO THE NEXT DAY.                                              SUB1 105
C                                                                      SUB1 106
         IF(METABL(I2,K).GE.9999.) GO TO 12                           SUB1 107
         I0=I-1                                                        SUB1 108
         METABL(I,K)=METABL(I2,K)-(METABL(I2,K)-METABL(I0,K))*.6667    SUB1 109
         METABL(I1,K)=METABL(I2,K)-(METABL(I2,K)-METABL(I0,K))*.3333   SUB1 110
   10 CONTINUE                                                         SUB1 111
C                                                                      SUB1 112
C        GENERATE SOLAR RADIATION TERM.                                SUB1 113
C                                                                      SUB1 114
         CALL SOLAR(LAT,IYR,IMON,IDAY,SRAD)                           SUB1 115
C                                                                      SUB1 116
C        APPLY CLOUD COVER ADJUSTMENT (AFTER WUNDERLICH) AND READ SOLAR RAD SUB1 117
C        IATION TERM INTO METABL.                                      SUB1 118
                                                                       SUB1 119
                                                                       SUB1 120
                                                                       SUB1 121
C                                                                      SUB1 122
         DO 13 I=1,25                                                  SUB1 123
   13 METABL(I,8)=SRAD(I)*.94*(1.-.65*METABL(I,8)**2)                  SUB1 124
C                                                                      SUB1 125
C        WRITE ONE DAY'S WEATHER RECORD IN TO INTERMEDIATE STORAGE.    SUB1 126
C                                                                      SUB1 127
         WRITE(9) METABL                                               SUB1 128
C                                                                      SUB1 129
C        IF NEXT DAY IS FIRST OF OCTOBER,SKIP TO NEXT MAY FIRST.       SUB1 130
C                                                                      SUB1 131
   20 IF(METABL(26,2).LE.9) GO TO 14                                   SUB1 132
C                                                                      SUB1 133
C        SEPARATE YEARS BY BLANK DATA RECORD.                          SUB1 134
C                                                                      SUB1 135
         DO 15 I=1,27                                                  SUB1 136
         DO 15 J=1,10                                                  SUB1 137
   15 METABL(I,J)=0.                                                   SUB1 138
         WRITE(9) METABL                                               SUB1 139
         DO 16 I=1,847                                                 SUB1 140
         READ(8)                                                       SUB1 141
C                                                                      SUB1 142
C        IF END OF RECORD ENCOUNTERED,RETURN TO MAIN PROGRAM.          SUB1 143
C                                                                      SUB1 144
         IF(EOF(8).NE.0) GO TO 17                                      SUB1 145
   16 CONTINUE                                                         SUB1 146
         GO TO 3                                                       SUB1 147
C                                                                      SUB1 148
C        READ IN NEXT DAY'S DATA.                                      SUB1 149
C                                                                      SUB1 150
   14 DO 18 I=1,3                                                      SUB1 151
         I24=I+24                                                      SUB1 152
         DO 18 K=1,10                                                  SUB1 153
   18 METABL(I,K)=METABL(I24,K)                                        SUB1 154
         METABL(1,4)=0.                                                SUB1 155
         DO 19 I=4,6                                                   SUB1 156
         METABL(I,1)=IDATE(1)                                          SUB1 157
         METABL(I,2)=IDATE(2)                                          SUB1 158
         METABL(I,3)=IDATE(3)                                          SUB1 159
         METABL(I,4)=IHOUR(I)                                          SUB1 160
         METABL(I,5)=WINDSP(I)                                         SUB1 161
         METABL(I,6)=TEMPDB(I)                                         SUB1 162
         METABL(I,7)=TEMPDP(I)                                         SUB1 163
         METABL(I,8)=SKY(I)                                            SUB1 164
         METABL(I,9)=SKY(I)                                            SUB1 165
   19 METABL(I,10)=HUMID(I)                                            SUB1 166
         INDEX=1                                                       SUB1 167
```

Figure B.1 (Continued).

```
      IYR=IDATE(1)                                                      SUB1  168
      IMON=IDATE(2)                                                     SUB1  169
      IDAY=IDATE(3)                                                     SUB1  170
      I=1                                                               SUB1  171
      GO TO 6                                                           SUB1  172
C                                                                       SUBI  173
C     WRITE ERROR MESSAGE WHEN DATA ARE SKIPPED                         SUB1  174
C                                                                       SUB1  175
   12 WRITE(6,500) IMON,IDAY,IYR                                        SUB1  176
  500 FORMAT(T35,'DISCONTINUITY IN DATA CAUSED ',I2,'/',I2,'/',I2,' TO BSUB1' 177
     1E SKIPPED')                                                       SUB1  178
C                                                                       .SUB1  179
C     FLAG RECORD CONTAINING BAD DATA.                                  SUB1  180
C                                                                       SUB1  181
      METABL(2,1)=9999.                                                 SUB1  182
      WRITE(9) METABL                                                   SUB1  183
      GO TO (3,20),INDEX                                                SUB1  184
   17 REWIND 9                                                          SUB1  185
      REWIND 8                                                          SUB1  186
      RETURN                                                            SUB1  187
      END                                                               SUB1  188
      SUBROUTINE SUB2(A,V,YRMAX)                                        JULY9   4
C   IMPROVED VERSION OF NUTTLE PROGRAM USING 2ND ORDER RK               SUB2    3
C   R CODELL,SEPT 19,1979                                               SUB2    4
C                                                                       SUB2    5
C     MODELS POND TEMPERATURE RESPONSE USING DATA IN INTERMEDIATE       SUB2    6
C     STORAGE.  RETURNS YEARLY MAXIMUM TEMPERATURES AND 30 DAY EVAPOR-  SUB2    7
C     ATIVE LOSSES WITH THEIR DATES OF OCCURENCE.                       SUB2    8
C                                                                       SUB2    9
      COMMON/TFUNC/ CON1,CON2                                           SUB2   10
      REAL ABSMAX(4),METABL(27,10),TIME(25),SRAD(25),TEMPDB(25),        SUB2   11
     1TEMPDP(25),WINDSP(25),KN(4),EV(4),EVAP(30),TEMPMX(5)              SUB2   12
     2,EVPMAX(4),YRMAX(40,8),MAXT                                       JULY9   5
      COMMON/COEF/ CEH(6),CEL(6),CH(6),CL(6)                            SUB2   14
      DATA TSTEP/1.0/                                                   SUB2   15
      DATA STEP/.5/                                                     SUB2   16
      DATA DTO2,DTO6,DT/.5,.16666667,1.0/                               SUB2   17
      DO 39 I=1,40                                                      SUB2   18
      DO 39 J=1,8                                                       SUB2   19
   39 YRMAX(I,J)=0.                                                     SUB2   20
      CON1=A/(1498*V)                                                   SUB2   21
      CON2=A/1497600.                                                   SUB2   22
      LNDX=0                                                            SUB2   23
      MAXT=0.                                                           JULY9   6
      ABSMAX(1)=0.                                                      SUB2   29
      EVPMAX(1)=0.                                                      SUB2   30
      TEMPMX(1)=0.                                                      SUB2   31
      EVTOT=0.                                                          SUB2   32
   10 READ(9) METABL                                                    SUB2   33
      IF(EOF(9).NE.0) GO TO 12                                          SUB2   34
      IF(METABL(2,1).GE.9999.) GO TO 10                                 SUB2   35
      PONDTP=METABL(1,7)                                                SUB2   36
      DO 30 I=1,30                                                      SUB2   37
   30 EVAP(I)=0                                                         SUB2   38
    1 CONTINUE                                                          SUB2   39
      DO 131 J=1,25                                                     SUB2   40
      SRAD(J)=METABL(J,8)                                               SUB2   41
      TEMPDB(J)=METABL(J,6)                                             SUB2   42
      TEMPDP(J)=METABL(J,7)                                             SUB2   43
      WINDSP(J)=METABL(J,5)                                             SUB2   44
  131 CONTINUE                                                          SUB2   45
      DO 132 J=1,24                                                     SUB2   46
      JP1=J+1                                                           SUB2   47
```

Figure B.1 (Continued).

```
C                                                                        SUB2  48
C      CALCULATION OF POND TEMPERATURE AND EVAPORATIVE WATER LOSS USING   SUB2  49
C      THE LINEAR HEAT EXCHANGE EQUATIONS IN A SECOND ORDER RUNGE-KUTTA   SUB2  50
C      NUMERICAL INTEGRATION.                                            SUB2  51
C                                                                        SUB2  52
       CALL TFUN(PONDTP,TEMPDB(J),WINDSP(J),SRAD(J),TEMPDP(J),           SUB2  53
      1 KN(1),EV(1))                                                     SUB2  54
       PTP1=PONDTP+KN(1)*DT                                              SUB2  55
       CALL TFUN(PTP1,TEMPDB(JP1),WINDSP(JP1),SRAD(JP1),TEMPDP(JP1),     SUB2  56
      1 KN(2),EV(2))                                                     SUB2  57
       PONDTP=PONDTP+(KN(1)+KN(2))*DTO2                                  SUB2  58
       EVAP(1)=EVAP(1)+(EV(1)+EV(2))*DTO2                                SUB2  59
C                                                                        SUB2. 60
C      COLLECT MAXIMUM TEMPERATURE                                       SUB2  61
C                                                                        SUB2  62
       IF(PONDTP.GT.MAXT) MAXT=PONDTP                                    JULY9  7
  132 CONTINUE                                                           SUB2  64
C                                                                        SUB2  65
C      SEARCH FOR YEARLY MAXIMUM TEMPERATURE AND EVAPORATIVE WATER LOSS. SUB2  66
C                                                                        SUB2  67
       DO 33 I=1,30                                                      SUB2  68
   33 EVTOT=EVTOT+EVAP(I)                                                SUB2  69
       IF(EVTOT.LT.EVPMAX(1))GO TO 13                                    SUB2  70
       EVPMAX(1)=EVTOT                                                   SUB2  71
       EVPMAX(2)=METABL(1,1)                                             SUB2  72
       EVPMAX(3)=METABL(1,2)                                             SUB2  73
       EVPMAX(4)=METABL(1,3)                                             SUB2  74
   13 DO 29 I=1,29                                                       SUB2  75
       I30=30-I                                                          SUB2  76
       I1=I30+1                                                          SUB2  77
   29 EVAP(I1)=EVAP(I30)                                                 SUB2  78
       EVAP(1)=0.                                                        SUB2  79
       EVTOT=0.                                                          SUB2  80
       IF(MAXT.LT.ABSMAX(1)) GO TO 8                                     JULY9  8
        ABSMAX(1)=MAXT                                                   JULY9  9
       ABSMAX(2)=METABL(1,1)                                             SUB2  83
       ABSMAX(3)=METABL(1,2)                                             SUB2  84
       ABSMAX(4)=METABL(1,3)                                             SUB2  85
    8 CONTINUE                                                           JULY9 10
       MAXT=0.                                                           JULY9 11
C                                                                        SUB2  90
C      READ IN NEXT DAY'S DATA.                                          SUB2  91
C                                                                        SUB2  92
   11 READ(9) METABL                                                     SUB2  93
       IF(EOF(9).NE.0.0) GOTO 12                                         SUB2  94
       IF(METABL(1,1).GT.0.) GO TO 14                                    SUB2  95
       LNDX=LNDX+1                                                       SUB2  96
       YRMAX(LNDX,1)=ABSMAX(1)                                           SUB2  97
       YRMAX(LNDX,2)=ABSMAX(2)                                           SUB2  98
       YRMAX(LNDX,3)=ABSMAX(3)                                           SUB2  99
       YRMAX(LNDX,4)=ABSMAX(4)                                           SUB2 100
       YRMAX(LNDX,5)=EVPMAX(1)                                           SUB2 101
       YRMAX(LNDX,6)=EVPMAX(2)                                           SUB2 102
       YRMAX(LNDX,7)=EVPMAX(3)                                           SUB2 103
       YRMAX(LNDX,8)=EVPMAX(4)                                           SUB2 104
       I=1                                                               SUB2 105
       IF(ABSMAX(1).GE.TEMPMX(1))GO TO 16                                SUB2 106
       I=6                                                               SUB2 107
       GO TO 20                                                          SUB2 108
   16 IF(I.GE.5) GO TO 17                                                SUB2 109
       I5=5-I                                                            SUB2 110
       DO 18 J=1,I5                                                      SUB2 111
       L=5-J                                                             SUB2 112
```

Figure B.1 (Continued).

```
      L1=L+1                                               SUB2 113
   18 TEMPMX(L1)=TEMPMX(L)                                 SUB2 114
   17 TEMPMX(I)=ABSMAX(1)                                  SUB2 117
   20 ABSMAX(1)=0.                                         SUB2 120
      EVPMAX(1)=0.                                         SUB2 121
      MAXT=0.0                                             JULY9 12
      GO TO 10                                             SUB2 124
   14 IF(METABL(2,1).LT.9999.) GO TO 1                     SUB2 125
      DO 37 I=1,35                                         SUB2 126
      I1=I+1                                               SUB2 127
   37 CONTINUE                                             JULY9 13
      GO TO 11                                             SUB2 130
C                                                          SUB2 131
C     END OF DATA FILE ENCOUNTERED. RETURN TO MAIN PROGRAM. SUB2 132
C                                                          SUB2 133
   12 REWIND 9                                             SUB2 134
      RETURN                                               SUB2 135
      END                                                  SUB2 136
      SUBROUTINE TFUN(PT,DB,W,SRAD,DP,DT,DE)               TFUN   2
      COMMON/TFUNC/ CON1,CON2                              TFUN   3
      DATA HSPRAY,HIN,ESPRAY/3*0.0/                        TFUN   4
      TSTAR=(DP+PT)*.5                                     TFUN   5
      BETA=.255-.0085*TSTAR+.000204*TSTAR**2               TFUN   6
      WINFUN=70+.7*W**2                                    TFUN   7
      RK=15.7+(.26+BETA)*WINFUN                            TFUN   8
      E=SRAD/RK+(.26*DB+BETA*DP)/(.26+BETA)                TFUN   9
      DT=(RK*(E-PT)+HIN-HSPRAY)*CON1                       TFUN  10
      DE=BETA*(PT-DP)*WINFUN*CON2-ESPRAY                   TFUN  11
      RETURN                                               TFUN  12
      END                                                  TFUN  13
      SUBROUTINE SUB3(YRMODY,IPRNT)                        JULY9 14
C                                                          JULY9 15
C     PRINTS AND/OR PRNCHES DATA FROM INTERMEDIATE         JULY9 16
C     FILE FOR PERIOD OF 'NDYS' DAYS BEFORE AND 5          JULY9 17
C     DAYS FOLLOWING YRMODY.                               JULY9 18
C                                                          JULY9 19
C         IF IPRINT=1,DATA IS PRINTED                      JULY9 20
C         IF IPRINT=-1, DATA IS PUNCHED                    JULY9 21
C         IF IPRINT=0, DATA IS BOTH PRINTED AND PUNCHED    JULY9 22
C                                                          JULY9 23
      REAL YRMODY(3),METABL(27,10),JNDX                    JULY9 24
      INTEGER IDATE(3)                                     JULY9 25
      N=0                                                  JULY9 26
      DATA NDYS/35/                                        JULY9 27
      JNDX=0.                                              JULY9 28
      IPNCH=0                                              JULY9 29
      IF(IPRNT.EQ.1) GO TO 40                              JULY9 30
      IF(IPRNT.EQ.0)IPRNT=1                                JULY9 31
      IPNCH=1                                              JULY9 32
   40 CONTINUE                                             JULY9 33
C                                                          JULY9 34
C     POSITION TAPE9 TO 'NDYS' DAYS BEFORE DATE            JULY9 35
C     PROVIDED IN YRMODY.  IF DATA IS NOT AVAILABLE,       JULY9 36
C     POSITION TAPE9 TO FIRST DAY OF DATA IN THE           JULY9 37
C     SAME YEAR AS YRMODY.                                 JULY9 38
C                                                          JULY9 39
      READ(9) METABL                                       JULY9 40
      YR=METABL(1,1)                                       JULY9 41
      REWIND 9                                             JULY9 42
      IF (YRMODY(1).LE.YR) GO TO 1                         JULY9 43
      N=(YRMODY(1)-YR)*154.                                JULY9 44
      DO 2 I=1,N                                           JULY9 45
    2 READ(9) METABL                                       JULY9 46
```

Figure B.1 (Continued).

```
      N=0                                                              JULY9 47
    1 IF (YRMODY(2).LE.5.)GO TO 3                                      JULY9 48
      N=((YRMODY(2)-5.)*31.)                                          JULY9 49
      IF(YRMODY(2).GT.6.)N=N-1                                        JULY9 50
    3 CONTINUE                                                         JULY9 51
      N=YRMODY(3)+N-NDYS                                              JULY9 52
      IF(N.GT.0)GO TO 4                                               JULY9 53
      NDYS=NDYS+N                                                     JULY9 54
      GO TO 6                                                          JULY9 55
    4 DO 5 I=1,N                                                      JULY9 56
    5 READ(9) METABL                                                  JULY9 57
    6 CONTINUE                                                         JULY9 58
      NDYS6=NDYS+6                                                    JULY9 59
      N=0                                                              JULY9 60
C                                                                      JULY9 61
C     GENERATE OUTPUT                                                  JULY9 62
C                                                                      JULY9 63
      DO 35 I=1,NDYS6                                                 JULY9 64
      READ(9)METABL                                                   JULY9 65
      IF(METABL(2,1).GE.9999.)GO TO 35                                JULY9 66
      IF(IPNCH.NE.1) GO TO 41                                         JULY9 67
      IF(I.EQ.1) PUNCH(4,610)NDYS6,METABL(1,2),METABL(1,3),METABL(1,1) JULY9 68
  610 FORMAT('** APPROXIMATELY ',I2,' DAYS OF MET. DATA FOLLOW. DATA AREJULY9 69
     1 PUNCHED 2 HOURS TO A',/,'**** CARD BEGINNING WITH HOUR 0 ON ',3F3SUB3 43
     2.0,'   THE FORMAT FOR THE DATA IS I3,2(',/,'**** 3F5.1,F6.1,F4.2,F4SUB3 44
     2.0) WHERE FIELD 1 IS THE CARD NUMBER AND THE FOLLOWING',/,'**** VASUB3 45
     3RIABLE SEQUENCE IS REPEATED- WIND SPEED,DRY BULB,DEWPOINT,SOLAR RASUB3 46
     50-',/,'**** IATION, CLOUD COVER,AND RELATIVE HUMIDITY.')         SUB3 47
      DO 42 L=1,23,2                                                   SUB3 48
      L1=L+1                                                           SUB3 49
      N=N+1                                                            SUB3 50
   42 WRITE(4,590)N,((METABL(J,K),K=5,10),J=L,L1)                      SUB3 51
  590 FORMAT (I3,2(3F5.1,F6.1,F4.2,F4.0))                             SUB3 52
      IF(IPRNT.NE.1) GO TO 35                                          SUB3 53
   41 CONTINUE                                                         SUB3 54
      IDATE(1)=METABL(1,2)                                            SUB3 55
      IDATE(2)=METABL(1,3)                                            SUB3 56
      IDATE(3)=METABL(1,1)                                            SUB3 57
      WRITE(6,500) IDATE                                              SUB3 58
      DO 39 J=1,24                                                    SUB3 59
   39 WRITE(6,520)(METABL(J,K),K=4,10)                                SUB3 60
      WRITE(6,510)                                                    SUB3 61
  500 FORMAT(1H1,5(/),T20,10('*'),' METEOROLOGY FOR '2(I2,'/'),I2,44('*'SUB3 62
     1),///,T25,71('.'),//,T25,',   HOUR   , WIND SP.,DRY BULB ,DEWPOINT ,SUB3 63
     2SOLAR RAD   CLOUD  ,RELATIVE ,',//,T25,',',T35,',  (MPH)  , (DEG.F)SUB3 64
     3 , (DEG.F) ,BTU/FT2/D,  COVER  ,HUMIDITY ,',//,T25,71('.'))       SUB3 65
  510 FORMAT(T25,71('.'))                                              SUB3 66
  520 FORMAT(T25,',',3X,F3.0,3X,',',2X,F4.1,3X,',',2X,F5.1,2X,',',2X,   SUB3 67
     1F5.1,2X,',',2X,F6.1,1X,',',3X,F4.2,2X,',',2X,F5.1,2X,',')         SUB3 68
   35 CONTINUE                                                         SUB3 69
   20 CONTINUE                                                         SUB3 100
      IF(IPNCH.EQ.1) WRITE(6,600)N                                    SUB3 101
  600 FORMAT(1H1,5(/),T20,10('*'),' UMBER OF CARDS PUNCHED = ',I3,' ',  SUB3 102
     140('*'))                                                         SUB3 103
      REWIND 9                                                         SUB3 104
      RETURN                                                           SUB3 105
      END                                                              JULY9 70
      SUBROUTINE SUB4(YRMODY,LAT)                                      SUB4   2
C                                                                      SUB4   3
C     PRINTS OUT AVERAGE MONTHLY VALUES FOR METEOROLOGIC PARAMETERS     SUB4   4
C     BEGINNING WITH DATE GIVEN IN YRMODY AND ENDING WITH THE LAST      SUB4   5
C     DAY ON THE DATA TAPE.                                           SUB4   6
C                                                                      SUB4   7
```

Figure B.1 (Continued).

```
      REAL YRMODY(3),METABL(27,10),LAT                              SUB4   8
      INTEGER IDATE(3),MON(5),MONTH(5)                              SUB4   9
      DATA MON/121,152,182,213,244/                                SUB4  10
      DATA MONTH/'MAY','JUNE','JULY','AUGUST','SEPTEMBER'/          SUB4  11
      INDX=0                                                        SUB4  12
      WINDSP=0.                                                     SUB4  13
      TEMPDP=0.                                                     SUB4  14
      TEMPDB=0.                                                     SUB4  15
      SOLARD=0.                                                     SUB4  16
      IDATE(1)=YRMODY(2)                                            SUB4  17
      CLOUD=0.                                                      SUB4  18
      HUMID=0.                                                      SUB4  19
      IDATE(2)=YRMODY(3)                                            SUB4  20
      IDATE(3)=YRMODY(1)                                            SUB4  21
      WRITE(6,500) IDATE                                            SUB4  22
  500 FORMAT(     5(/),T20,10('*'),' THE MONTHLY AVERAGE VALUES FROM', SUB4  23
     12(I2,'/'),I2,' TO END OF DATA ',13('*'),//)                  SUB4  24
      WRITE(6,510)                                                  SUB4  25
  510 FORMAT(T30,61('.'),/,T30,'*RMS WIND *DRY BULB *DEWPOINT *  SOLAR  *SUB4  26
     1  CLOUD  *RELATIVE *',/,T30,'*  SPEED  * (DEG.F) * (DEG.F) *RADIATSUB4  27
     2ION*  COVER  *HUMIDITY *')                                   SUB4  28
      IYR=1900+IDATE(3)                                             SUB4  29
      WRITE(6,520) IYR                                              SUB4  30
  520 FORMAT(T20,I4,T30,61('.'),/,T30,'*',T40,'*',T50,'*',T60,'*',T70, SUB4  31
     1'*',T80,'*',T90,'*')                                         SUB4  32
C                                                                  SUB4  33
C     POSITION TAPE9 TO FIRST DAY OF MONTH PROVIDED IN YRMODY.      SUB4  34
C                                                                  SUB4  35
      READ(9) METABL                                               SUB4  36
      YR=METABL(1,1)                                                SUB4  37
      REWIND 9                                                      SUB4  38
      IF(YRMODY(1).LE.YR) GO TO 1                                   SUB4  39
      N=(YRMODY(1)-YR)*154.+1.                                      SUB4  40
      DO 2 I=1,N                                                    SUB4  41
    2 READ(9)METABL                                                SUB4  42
    1 N=((YRMODY(2)-5.)*31.)                                        SUB4  43
      IF(N.LE.0) GO TO 6                                            SUB4  44
      DO 4 I=1,N                                                    SUB4  45
    4 READ(9) METABL                                                SUB4  46
    6 IF(METABL(1,3).LE.1.) GO TO 5                                 SUB4  47
      BACKSPACE 9                                                   SUB4  48
      READ(9)METABL                                                 SUB4  49
      GO TO 6                                                       SUB4  50
    5 IF(METABL(2,1).GE.9999.) GO TO 9                              SUB4  51
C                                                                  SUB4  52
C     READ IN ONE MONTH'S DATA                                     SUB4  53
C                                                                  SUB4  54
    8 INDX=INDX+1                                                   SUB4  55
      IDATE(1)=METABL(1,2)                                          SUB4  56
      IDATE(2)=METABL(1,3)                                          SUB4  57
      IDATE(3)=METABL(1,1)                                          SUB4  58
      DAYNUM=MON(IDATE(1)-4)+IDATE(2)-1                             SUB4  59
      IF(MOD(IDATE(3),4).EQ.0) DAYNUM=DAYNUM+1.                     SUB4  60
      DAYLEN=DAYLIT(LAT,DAYNUM)                                     SUB4  61
      DO 7 I=1,24                                                   SUB4  62
      WINDSP=METABL(I,5)**2+WINDSP                                  SUB4  63
      TEMPDB=METABL(I,6)+TEMPDB                                     SUB4  64
      TEMPDP=METABL(I,7)+TEMPDP                                     SUB4  65
      CLOUD=METABL(I,9)+CLOUD                                       SUB4  66
      HUMID=METABL(I,10)+HUMID                                      SUB4  67
    7 SOLARD=SOLARD+METABL(I,8)/DAYLEN                              SUB4  68
    9 READ (9) METABL                                              SUB4  69
      IF(METABL(1,1).LE.0.) GO TO 11                                SUB4  70
```

Figure B.1 (Continued).

91

```
      IF(METABL(1,3).LE.1.) GO TO 10                                   SUB4   71
      IF(METABL(2,1).GE.9999.) GO TO 9                                 SUB4   72
      GO TO 8                                                          SUB4   73
   10 DAYS=INDX                                                        SUB4   74
C                                                                      SUB4   75
C     CALCULATE AND PRINT AVERAGES                                     SUB4   76
C                                                                      SUB4   77
      INDX=0                                                           SUB4   78
      AVGWS=(WINDSP/DAYS/24.)**.5                                      SUB4   79
      AVGDP=TEMPDP/DAYS /24.                                           SUB4   80
      AVGDB=TEMPDB/DAYS/24.                                            SUB4   81
      AVGCL=CLOUD/DAYS/24.                                             SUB4   82
      AVGHM=HUMID/DAYS/24.                                             SUB4   83
      AVGSR=SOLARD/DAYS                                                SUB4   84
      I=IDATE(1)-4                                                     SUB4   85
      WRITE(6,530)MONTH(I),AVGWS,AVGDB,AVGDP,AVGSR,AVGCL,AVGHM         SUB4   86
  530 FORMAT(T20,A10,'*',2X,F5.2,2X,'*',2X,F5.2,2X,'*',2X,F5.2,2X,'*',1XSUB4  87
     1,F6.1,2X,'*',2X,F4.2,3X,'*',1X,F5.1,3X,'*',/,T30,'*',T40,'*',T50, SUB4  88
     3'*',T60,'*',T70,'*',T80,'*',T90,'*')                             SUB4  89
      WINDSP=0.                                                        SUB4   90
      TEMPDP=0.                                                        SUB4   91
      TEMPDB=0.                                                        SUB4   92
      CLOUD=0.                                                         SUB4   93
      HUMID=0.                                                         SUB4   94
      SOLARD=0.                                                        SUB4   95
      GO TO 5                                                          SUB4   96
   11 DAYS=INDX                                                        SUB4   97
C                                                                      SUB4   98
C     CALCULATE AND PRINT AVERAGES FOR THE LAST MONTH OF EACH DATA     SUB4   99
C     PERIOD                                                           SUB4  100
C                                                                      SUB4  101
      INDX=0                                                           SUB4  102
      AVGWS=(WINDSP/DAYS/24.)**.5                                      SUB4  103
      AVGDP=TEMPDP/DAYS/24.                                            SUB4  104
      AVGDB=TEMPDB/DAYS/24.                                            SUB4  105
      AVGCL=CLOUD/DAYS/24.                                             SUB4  106
      AVGHM=HUMID/DAYS/24.                                             SUB4  107
      AVGSR=SOLARD/DAYS                                                SUB4  108
      WRITE(6,530) MONTH(5),AVGWS,AVGDB,AVGDP,AVGSR,AVGCL,AVGHM        SUB4  109
      WINDSP=0.                                                        SUB4  110
      TEMPDP=0.                                                        SUB4  111
      TEMPDB=0.                                                        SUB4  112
      SOLARD=0.                                                        SUB4  113
      CLOUD=0.                                                         SUB4  114
      HUMID=0.                                                         SUB4  115
      READ (9) METABL                                                  SUB4  116
      IF(EOF(9).NE.0) GO TO 12                                         SUB4  117
      IYR=1900+METABL(1,1)                                             SUB4  118
      WRITE(6,520) IYR                                                 SUB4  119
      IF(METABL(2,1).GE.9999.) GO TO 9                                 SUB4  120
      GO TO 8                                                          SUB4  121
   12 WRITE(6,540)                                                     SUB4  122
  540 FORMAT(T30,61('.'))                                             SUB4  123
      RETURN                                                           SUB4  124
      END                                                              SUB4  125
      SUBROUTINE SUB5(YRMAX)                                           SUB5    2
C                                                                      SUB5    3
C     COMPUTES SAMPLE MEAN, STANDARD DEVIATION,SKEW, AND EXCEEDENCE FOR SUB5   4
C     YEARLY MAXIMUM TEMPERATURES AND WATER LOSSES GENERATED BY SUB2   SUB5    5
C                                                                      SUB5    6
      REAL YRMAX(40,8),JUNK(4),P(40),MT,ME                            SUB5    7
      SUMT=0.                                                          SUB5    8
      SUMT2=0.                                                         SUB5    9
      SUMT3=0.                                                         SUB5   10
```

Figure B.1 (Continued).

92

```
      SUME=0.                                                       SUBS  11
      SUME2=0.                                                      SUBS  12
      SUME3=0.                                                      SUBS  13
      DO 20 L=1,40                                                  SUBS  14
      IF(YRMAX(L,1).LE.0.) GO TO 21                                 SUBS  15
   20 CONTINUE                                                      SUBS  16
      L=L+1                                                         SUBS  17
   21 L=L-1                                                         SUBS  18
C                                                                   SUBS  19
C     RANK DATA IN ORDER OF DECREASING MAGNITUDE                    SUBS  20
C                                                                   SUBS  21
      DO 1 J=1,5,4                                                  SUBS  22
      DO 1 I=2,L                                                    SUBS  23
      I1=I-1                                                        SUBS  24
      IF(YRMAX(I,J).LE.YRMAX(I1,J)) GO TO 1                         SUBS  25
      DO 2 M=1,4                                                    SUBS  26
      MJ=M+J-1                                                      SUBS  27
    2 JUNK(M)=YRMAX(I,MJ)                                           SUBS  28
      DO 3 M=1,I                                                    SUBS  29
      IF(JUNK(1).GT.YRMAX(M,J)) GO TO 4                             SUBS  30
    3 CONTINUE                                                      SUBS  31
    4 DO 5 K=M,I1                                                   SUBS  32
      KM=I-K+M                                                      SUBS  33
      KM1=KM-1                                                      SUBS  34
      DO 5 L2=1,4                                                   SUBS  35
      LJ=L2+J-1                                                     SUBS  36
    5 YRMAX(KM,LJ)=YRMAX(KM1,LJ)                                    SUBS  37
      DO 6 L2=1,4                                                   SUBS  38
      LJ=L2+J-1                                                     SUBS  39
    6 YRMAX(M,LJ)=JUNK(L2)                                          SUBS  40
    1 CONTINUE                                                      SUBS  41
C                                                                   SUBS  42
C     COMPUTE EXCEEDENCES                                           SUBS  43
C                                                                   SUBS  44
      RL=L                                                          SUBS  45
      P(1)=(1.-(.5)**(1./RL))*100.                                  SUBS  46
      X=2.*(50.-P(1))/(RL-1.)                                       SUBS  47
      DO 7 I=2,L                                                    SUBS  48
      I1=I-1                                                        SUBS  49
    7 P(I)=P(I1)+X                                                  SUBS  50
      DO 22 I=1,L                                                   SUBS  51
      SUMT=SUMT+YRMAX(I,1)                                          SUBS  52
      SUMT2=SUMT2+YRMAX(I,1)**2                                     SUBS  53
      SUMT3=SUMT3+YRMAX(I,1)**3                                     SUBS  54
      SUME=SUME+YRMAX(I,5)                                          SUBS  55
      SUME2=SUME2+YRMAX(I,5)**2                                     SUBS  56
   22 SUME3=SUME3+YRMAX(I,5)**3                                     SUBS  57
      MT=SUMT/RL                                                    SUBS  58
      ST=SQRT((SUMT2-(SUMT**2/RL))/(RL-1.))                         SUBS  59
      GT=(RL**2*SUMT3-3.*RL*SUMT*SUMT2+2.*SUMT**3)/(ST**3*RL*(RL-1.)*SUBS  60
     1(RL-2.))                                                      SUBS  61
      ME=SUME/RL                                                    SUBS  62
      SE=SQRT((SUME2-(SUME**2/RL))/(RL-1.))                         SUBS  63
      GE=(RL**2*SUME3-3.*RL*SUME*SUME2+2.*SUME**3)/(SE**3*RL*(RL-1.)*SUBS  64
     1(RL-2.))                                                      SUBS  65
      WRITE(6,530)                                                  SUBS  66
  530 FORMAT(////)                                                  SUBS  67
      WRITE(6,500)                                                  SUBS  68
  500 FORMAT(T20,10('*'),'THE SAMPLE OF YEARLY MAXIMUM POND TEMPERATURESSUBS  69
     1 AND 30 DAY ', 10('*'),/,T31,'EVAPORATIVE LOSSES GENERATED BY THISUBS  70
     2S MODEL IS DESCRIBED BELOW.',////,T28,10('.'),'TEMPERATURE',19('.')SUBS  71
     3,'EVAPORATIVE LOSS',9('.'),//,T28,'*EXCEEDED',15X, 'DATE  *EXCEESUBS  72
     4DED',15X,'DATE  *',/, T28,'*/100 YR*  (DEG.F)  *(YR.MO.DY.)*/100SUBS  73
     5 YR*  FT**3  *(YR.MO.DY.)*',/,T28,65('.'))                    SUBS  74
```

Figure B.1 (Continued).

93

```
         DO 10 I=1,L                                                   SUB5  75
      10 WRITE(6,510) P(I),(YRMAX(I,J),J=1,4),P(I),(YRMAX(I,K),K=5,8)    SUB5  76
     510 FORMAT(T28,'*',1X,F5.2,1X,'*',3X,F5.2,3X,'*',1X,3F3.0,1X,'*',1X,SUB5  77
        2F5.2,1X,'*',1X,F9.1,1X,'*',1X,3F3.0,1X,'*')                     SUB5  78
         WRITE(6,520) MT,ME,ST,SE,GT,GE                                 SUB5  79
     520 FORMAT(T28,65('.'),//,T26,'MEAN',T40,F5.2,T70,F9.1,//,T17,       SUB5  80
        1'STANDARD DEV.',T40,F6.3,T70,F10.2,//,T26,'SKEW',T40,F6.3,T70,   SUB5  81
        2F11.3)                                                          SUB5  82
         RETURN                                                         SUB5  83
         END                                                           SUB5  84
         SUBROUTINE READRC                                             READRC  2
C                                                                      READRC  3
C        READS WIND SPEED, DRY BULB TEMPERATURE, WET BULB TEMPERATURE, READRC  4
C        DEW POINT, RELATIVE HUMIDITY, STATION PRESSURE, AND TENTHS OF READRC  5
C        CLOUD COVER FROM NATIONAL WEATHER SERVICE DATA TAPES.  WIND SPEED READRC 6
C        IS RETURNED IN MPH, TEMPERATURE IN DEGREES FARENHEIT, AND PRESSUREREADRC  7
C        IN MM-HG.  INPUT RECORD IS 495 CHARACTERS LONG.               READRC  8
C                                                                      READRC  9
         INTEGER JUNK(6,9),ISTAT(2),IWIND(6,4),ITEMP(6,6),IHUMID(6,2), READRC 10
        1IPRESS(6,4),ISKY(6,6)                                          READRC 11
         COMMON IDATE(3), IHOUR(6),WINDSP(6),TEMPDB(6),TEMPWB(6),TEMPDP(6),READRC 12
        1HUMID(6),PRESSR(6),SKY(6)                                      READRC 13
         READ(8,500)    ISTAT,IDATE,(IHOUR(I),(JUNK(I,K),K=1,4),        READRC 14
        1(IWIND(I,K),K=1,4),(ITEMP(I,K),K=1,6),IHUMID(I,1),IHUMID(I,2), READRC 15
        2(IPRESS(I,K),K=1,4),(ISKY(I,K),K=1,6),(JUNK(I,K),K=5,9),I=1,6) READRC 16
     500 FORMAT ( · I4,I5,3I2,6(I2,1X,I2,A1,1X,I2,A1,I1,A1,4(I2,A1),1X, READRC 17
        1I2,A1,I4,A1,I3,A1,1X,6A1,   2(A10),A2,A8,A2,4X))               READRC 18
         DO 100 I=1,6                                                  READRC 19
         CALL SIGNCK (IWIND(I,3),IWIND(I,4))                           READRC 20
         WINDSP(I)=IWIND(I,3)                                          READRC 21
         CALL SIGNCK (ITEMP(I,1),ITEMP(I,2))                           READRC 22
         WINDSP(I)=WINDSP(I)*1.15078                                   READRC 23
         CALL SIGNCK (ITEMP(I,3),ITEMP(I,4))                           READRC 24
         CALL SIGNCK (ITEMP(I,5),ITEMP(I,6))                           READRC 25
         TEMPDB(I)=ITEMP(I,1)                                          READRC 26
         TEMPWB(I)=ITEMP(I,3)                                          READRC 27
         TEMPDP(I)=ITEMP(I,5)                                          READRC 28
         CALL SIGNCK (IHUMID(I,1),IHUMID(I,2))                         READRC 29
         HUMID(I)=IHUMID(I,1)                                          READRC 30
         CALL SIGNCK (IPRESS(I,3),IPRESS(I,4))                         READRC 31
         PRESSR(I)=IPRESS(I,3)                                         READRC 32
         PRESSR(I)=PRESSR(I)*.01                                       READRC 33
         ICOVER=0                                                      READRC 34
         CALL SIGNCK(ICOVER,ISKY(I,5))                                 READRC 35
     100 SKY(I)=ICOVER*.1                                              READRC 36
         RETURN                                                        READRC 37
         END                                                           READRC 38
         SUBROUTINE SIGNCK(IFLD,ISGN)                                  SIGNCK  2
C        THIS SUBROUTINE FURNISHED BY NATIONAL CLIMATIC CENTER, ASHEVILLE SIGNCK  3
C            WILL TEST ANY PSYCHROMETRIC WITH A SIGN-OVER-UNITS        SIGNCK  4
C            POSITION READ AS A1 AND THE HIGH ORDER POSITION AS AN     SIGNCK  5
C            I SPECIFICATION OF PROPER WIDTH                           SIGNCK  6
C        THE SIGN SHOULD ENTER THE PARAMETER LIST AS ISGN,            SIGNCK  7
C            THE REMAINING PORTION AS IFLD                             SIGNCK  8
C        UPON RETURN FROM THE SUBROUTINE THE VALUE OF IFLD WILL BE     SIGNCK  9
C            AN INTEGER WITH PROPER SIGN'                              SIGNCK 10
C        IT WILL BE THE USER'S RESPONSIBILITY TO CONVERT THIS          SIGNCK 11
C            TO  DECIMAL WITH PROPER DECIMAL ALIGNMENT                 SIGNCK 12
C        INVALID CONDITION CAUSES IFLD TO BE SET TO 9999               SIGNCK 13
         DIMENSION IP(10),MIN(10),NUM(10)                              SIGNCK 14
         DIMENSION INUM(10)                                            SIGNCK 15
         DATA INUM/'1','2','3','4','5','6','7','8','9','0'/             SIGNCK 16
C        NOTE - SOME COMPUTER SYSTEMS MAY REQUIRE DIFFERENT CHARACTERS AS SIGNCK 17
C        THE LAST CHARACTERS IN ARRAYS IP AND MIN                      SIGNCK 18
```

Figure B.1 (Continued).

94

```
      DATA MIN/'J','K','L','M','N','O','P','Q','R','1'/        SIGNCK19
      DATA IP/'A','B','C','D','E','F','G','H','I',             SIGNCK20
     1 7255555555555555555555558/                             SIGNCK21
      DATA NUM/1,2,3,4,5,6,7,8,9,0/                            SIGNCK22
      DATA IAST/'*'/                                           SIGNCK23
      DATA MINUS/'-'/                                          SIGNCK24
      DATA NULL/' '/                                           SIGNCK25
      IF (ISGN.EQ.NULL.AND.IFLD.NE.0) GO TO 125                SIGNCK26
      IF (ISGN.EQ.IAST) GO TO 105                              SIGNCK27
      IF (ISGN.EQ.MINUS) GO TO 110                             SIGNCK28
      DO 100 K=1,10                                            SIGNCK29
      IF (ISGN.EQ.IP(K)) GO TO 115                             SIGNCK30
      IF (ISGN.EQ.MIN(K)) GO TO 120                            SIGNCK31
      IF (ISGN.EQ.INUM(K)) GO TO 115                           SIGNCK32
  100 CONTINUE                                                 SIGNCK33
  105 IFLD=999999                                              JULY9 71
      RETURN                                                   SIGNCK35
  110 IFLD=10                                                  SIGNCK36
      RETURN                                                   SIGNCK37
  125 IFLD=IFLD*10                                             SIGNCK38
      RETURN                                                   SIGNCK39
  115 IFLD=IFLD*10+NUM(K)                                      SIGNCK40
      RETURN                                                   SIGNCK41
  120 IFLD=-(IFLD*10+NUM(K))                                   SIGNCK42
      RETURN                                                   SIGNCK43
      END                                                      SIGNCK44
      SUBROUTINE SOLAR (LAT,YR,MONTH,DAY,SRAD)                 SOLAR  2
C                                                              SOLAR  3
C     RETURNS INSOLATION IN BTU/FT**2/DAY AT EACH HOUR OF THE DAY.  SOLAR  4
C                                                              SOLAR  5
      INTEGER YR,MONTH,DAY,MONDAT(12)                          SOLAR  6
      REAL LAT,SRAD(25)                                        SOLAR  7
      DATA MONDAT/0,31,59,90,120,151,181,212,243,273,304,334/ SOLAR  8
      LP=MOD(YR,4)                                             SOLAR  9
      IF(LP.NE.0)GO TO 120                                     SOLAR 10
      DO 100 I=3,12                                            SOLAR 11
  100 MONDAT(I)=MONDAT(I)+1                                    SOLAR 12
  120 NUM=MONDAT(MONTH)+DAY                                    SOLAR 13
C                                                              SOLAR 14
C     FIND TOTAL POSSIBLE DAILY RADIATION AND LENGTH OF DAYLIGHT.  SOLAR 15
C                                                              SOLAR 16
      TOTRAD=HAMN(LAT,YR,NUM)                                  SOLAR 17
      DAYNUM=NUM                                               SOLAR 18
      DAYLEN=DAYLIT(LAT,DAYNUM)                                SOLAR 19
C                                                              SOLAR 20
C     CALCULATE THE SINUSOIDAL VARIATION IN DAILY RADIATION.   SOLAR 21
C                                                              SOLAR 22
      T1=.5*DAYLEN                                             SOLAR 23
      A=.5*TOTRAD                                              SOLAR 24
      W=.2618                                                  SOLAR 25
      ALPHA=1./(1./(A*W)*SIN(W*T1)-T1/A*COS(W*T1))             SOLAR 26
      ALPTO=ALPHA*COS(W*T1)                                    SOLAR 27
      DO 130 I=1,25                                            SOLAR 28
      TO=I-1.                                                  SOLAR 29
      SRAD(I)=0.                                               SOLAR 30
      TO=ABS(TO-12.)                                           SOLAR 31
C                                                              SOLAR 32
C     CALCULATE RATE OF INSOLATION FOR EACH HOUR OF DAYLIGHT.  SOLAR 33
C                                                              SOLAR 34
      IF(TO.LE.T1)SRAD(I)=(ALPHA*COS(W*TO)-ALPTO)*DAYLEN       SOLAR 35
  130 CONTINUE                                                 SOLAR 36
      IF(LP.NE.0) RETURN                                       SOLAR 37
      DO 140 I=3,12                                            SOLAR 38
  140 MONDAT(I)=MONDAT(I)-1                                    SOLAR 39
      RETURN                                                   SOLAR 40
      END                                                      SOLAR 41
```

Figure B.1 (Continued).

```
      FUNCTION DAYLIT(LAT,DAYNUM)                                       DAYLIT  2
C                                                                       DAYLIT  3
C     RETURNS HOURS OF DAYLIGHT GIVEN LATITUDE OF OBSERVATION AND       DAYLIT  4
C     NUMBER OF THE DAY OF THE YEAR.  LATITUDE MUST BE BETWEEN 25 AND   DAYLIT  5
C     50 DEGREES NORTH.  THE SOURCE FOR THE LENGTH OF DAYLIGHT INFOR-   DAYLIT  6
C     MATION (STORED IN ARRAY 'LENGTH') IS THE SMITHSONIAN METEOROLOG-  DAYLIT  7
C     ICAL TABLES.                                                      DAYLIT. 8
C                                                                       DAYLIT  9
      REAL LAT,LATBL(6),LENGTH(6,10) ,DAY(10)                           DAYLIT 10
      DATA LATBL/25.,30.,35.,40.,45.,50.01/                            DAYLIT 11
      DATA DAY/-10.,13.,79.,145.,172.,197.,263.,333.,355.,378./         DAYLIT 12
      DATA (LENGTH(1,I),I=1,10)                                         DAYLIT 13
     1 /10.58,10.73,12.15,13.50,13.68,13.53,12.17,10.73,10.58,10.73/    DAYLIT 14
      DATA (LENGTH(2,I),I=1,10)                                         DAYLIT 15
     2/10.20,10.40,12.15,13.83,14.08,13.87,12.17,10.40,10.20,10.40/     DAYLIT 16
      DATA (LENGTH(3,I),I=1,10)                                         DAYLIT 17
     3/9.80,10.03,12.15,14.23,14.52,14.26,12.20,10.02,9.80,10.03/       DAYLIT 18
      DATA (LENGTH(4,I),I=1,10)                                         DAYLIT 19
     4/9.33,9.60,12.18,14.67,15.02,14.70,12.22,9.60,9.33,9.60/          DAYLIT 20
      DATA (LENGTH(5,I),I=1,10)                                         DAYLIT 21
     5/8.75,9.10,12.19,15.28,15.61,15.23,12.23,9.09,8.75,9.10/          DAYLIT 22
      DATA (LENGTH(6,I),I=1,10)                                         DAYLIT 23
     6/8.07,8.50,12.22,15.83,16.38,15.88,12.28,8.48,8.07,8.50/          DAYLIT 24
      DO 100 I=2,10                                                     DAYLIT 25
      I1=I-1                                                            DAYLIT 26
      IF(DAYNUM.GE.DAY(I1).AND.DAYNUM.LT.DAY(I))GO TO 110               DAYLIT 27
  100 CONTINUE                                                          DAYLIT 28
  110 DO 120 K=2,6                                                      DAYLIT 29
      K1=K-1                                                            DAYLIT 30
      IF(LAT.GE.LATBL(K1).AND.LAT.LT.LATBL(K)) GO TO 130                DAYLIT 31
  120 CONTINUE                                                          DAYLIT 32
C                                                                       DAYLIT 33
C     LINEAR INTERPOLATION OF TABLE 'LENGTH'.                           DAYLIT 34
C                                                                       DAYLIT 35
  130 DELDY=(DAY(I)-DAYNUM)/(DAY(I)-DAY(I1))                            DAYLIT 36
      A=LENGTH(K1,I)-(DELDY*(LENGTH(K1,I)-LENGTH(K1,I1)))               DAYLIT 37
      B=LENGTH(K,I)-(DELDY*(LENGTH(K,I)-LENGTH(K,I1)))                  DAYLIT 38
      DAYLIT=B-(LATBL(K)-LAT)/5.*(B-A)                                  DAYLIT 39
      RETURN                                                            DAYLIT 40
      END                                                               DAYLIT 41
      FUNCTION HAMN(LAT,YR,MODA)                                        HAMN   2
C     SOLAR RADIATION ON HORIZONTAL SURFACE                             HAMN   3
C         FROM HAMON, WEISS, + WILSON    )100(>                         HAMN   4
C         #MONTHLY WEATHER REVIEW#--PAGE 141--JUNE 1954                 HAMN   5
C     PROGRAM AUTHOR--E.C.LONG.  COMPUTER SCIENCES DIVISION--ORNL       HAMN   6
C     UNION CARBIDE NUCLEAR DIVISION.  OAK RIDGE, TENNESSEE             HAMN   7
C **** DAILY RADIATION RETURNED IN BTU'S ****                          HAMN   8
      REAL DATE(16),L25(16),L30(16),L35(16),L40(16),L45(16),L50(16),    HAMN   9
     1     LT(6),LAT,X(3),Y(3),L(96)                                    HAMN  10
      INTEGER IM(12),N(12),YR                                          HAMN  11
      EQUIVALENCE (L(1),L25(1)),(L(17),L30(1)),(L(33),L35(1)),          HAMN  12
     1    (L(49),L40(1)),(L(65),L45(1)),(L(81),L50(1))                  HAMN  13
      DATA DATE    /-41.0,-11.0,20.0,51.0,79.0,110.0,140.0,             HAMN  14
     1    171.0,201.0,232.0,263.0,293.0,324.0,354.0,385.0,416.0/        HAMN  15
      DATA L25    /1754.0,1616.0,1794.0,2116.0,2399.0,2611.0,2708.0,    HAMN  16
     1    2729.0,2695.0,2571.0,2338.0,2030.0,1754.0,1616.0,             HAMN  17
     2    1794.0,2116.0/                                                HAMN  18
      DATA L30    /1557.0,1390.0,1570.0,1909.0,2266.0,2557.0,2699.0,    HAMN  19
     1    2729.0,2662.0,2503.0,2224.0,1873.0,1557.0,1390.0,             HAMN  20
     2    1570.0,1909.0/                                                HAMN  21
      DATA L35    /1338.0,1149.0,1351.0,1723.0,2124.0,2492.0,2680.0,    HAMN  22
     1    2729.0,2645.0,2426.0,2064.0,1685.0,1338.0,1149.0,             HAMN  23
     2    1351.0,1723.0/                                                HAMN  24
```

Figure B.1 (Continued).

96

```
      DATA L40    /1103.0,909.7,1103.0,1514.0,1947.0,2397.0,2655.0,       HAMN  25
     1    2729.0,2603.0,2342.0,1951.0,1479.0,1103.0,909.7,                 HAMN  26
     2'   1103.0,1514.0/                                                   HAMN  27
      DATA L45    /882.7,687.3,881.0,1311.0,1778.0,2289.0,2618.0,          HAMN  28
     1    2729.0,2571.0,2247.0,1769.0,1274.0,882.7,687.3,881.0,1311.0/     HAMN  29
      DATA L50    /682.3,463.3,631.0,1053.0,1568.0,2165.0,2581.0,          HAMN  30
     1    2729.0,2527.0,2136.0,1584.0,1060.0,682.3,463.3,631.7,1053.0/     HAMN  31
      DATA LT   /25.0,30.0,35.0,40.0,45.0,50.0/                            HAMN  32
      DATA    IM   /1,32,60,91,121,152,182,213,244,274,305,335/            HAMN  33
      DATA    N    /31,28,31,30,31,30,31,31,30,31,30,31/                   HAMN  34
      DAYC=MODA                                                            HAMN  35
      LEAP=MOD(YR,4)                                                       HAMN  36
      IF (LEAP.NE.0) GO TO 110                                             HAMN  37
      DO 100 I=4,16                                                        HAMN  38
      DATE(I)=DATE(I)+1.0                                                  HAMN  39
  100 CONTINUE                                                             HAMN  40
      DO 105 I=2,11                                                        HAMN  41
      IM(I)=IM(I)+1                                                        HAMN  42
      N(I)=N(I)+1                                                          HAMN  43
  105 CONTINUE                                                             HAMN  44
  110 SUM=0.0                                                              HAMN  45
      IF (MODA.GT.0) GO TO 115                                             HAMN  46
C     FOR MODA)0 FIND AVERAGE SOLAR RADIATION FOR MONTH -MODA              HAMN  47
      MO=-MODA                                                             HAMN  48
      I1=IM(MO)                                                            HAMN  49
      ID=N(MO)                                                             HAMN  50
      I2=I1+ID-1                                                           HAMN  51
      DAYS=ID                                                              HAMN  52
      DAY=I1                                                               HAMN  53
      GO TO 120                                                            HAMN  54
C     FOR MODA>0 FIND RADIATION FOR DAY #DAYC#                             HAMN  55
C         DAYC IS EQUIVALENCED TO MODA                                     HAMN  56
  115 I1=1                                                                 HAMN  57
      ID=1                                                                 HAMN  58
      I2=1                                                                 HAMN  59
      DAY=DAYC                                                             HAMN  60
      DAYS=1.0                                                             HAMN  61
  120 DO 180 II=I1,I2                                                      HAMN  62
C     DETERMINE IF DAY IS TABULAR                                          HAMN  63
C         OF IF DAY NOT TABULAR, INDEX OF DAY                              HAMN  64
      MD=0                                                                 HAMN  65
      MI=0                                                                 HAMN  66
      DO 130 I=2,14                                                        HAMN  67
      DATEI=DATE(I)                                                        HAMN  68
      IF (DAY.NE.DATEI) GO TO 125                                          HAMN  69
      MD=I                                                                 HAMN  70
      GO TO 140                                                            HAMN  71
C     MD HAS INDEX I IF DAY=DATE(I)                                        HAMN  72
  125 IF (DAY.GT.DATEI.AND.DAY.LT.DATE(I+1)) GO TO 135                     HAMN  73
  130 CONTINUE                                                             HAMN  74
      GO TO 140                                                            HAMN  75
  135 MI=I                                                                 HAMN  76
C     MI=I FOR DATE(I))DAY)DATE(I+1)                                       HAMN  77
C     DETERMINE IF LAT IS TABULAR VALUE                                    HAMN  78
  140 IF (MODA.LT.0.AND.II.GT.I1) GO TO 150                                HAMN  79
      ML=0                                                                 HAMN  80
      DO 145 I=1,6                                                         HAMN  81
      IF (LAT.NE.LT(I)) GO TO 145                                          HAMN  82
      ML=I                                                                 HAMN  83
C     ML=I FOR LAT TABULAR VALUE                                           HAMN  84
      GO TO 150                                                            HAMN  85
  145 CONTINUE                                                             HAMN  86
  150 IF (MD*ML.EQ.0) GO TO 155                                            HAMN  87
```

Figure B.1 (Continued).

```
C     TABULAR DATE + LATITUDE                               HAMN  88
      J=(ML-1)*16+MD                                        HAMN  89
      HAMN=L(J)                                             HAMN  90
      GO TO 175                                             HAMN  91
  155 IF (ML.EQ.0) GO TO 160                                HAMN  92
C     NON TABULAR DATE + TABULAR LATITUDE                   HAMN  93
      MI1=MI-1                                              HAMN  94
      J=(ML-1)*16+MI1                                       HAMN  95
      HAMN=YLAG(DAY,DATE(MI1),L(J),4)                       HAMN  96
      GO TO 175                                             HAMN  97
  160 IF (LAT.LE.32.5) LATF=1                               HAMN  98
      IF (LAT.GT.32.5.AND.LAT.LE.37.5) LATF=2               HAMN  99
      IF (LAT.GT.37.5.AND.LAT.LE.42.5) LATF=3               HAMN 100
      IF (LAT.GT.42.5) LATF=4                               HAMN 101
      X(1)=LT(LATF)                                         HAMN 102
      X(2)=LT(LATF+1)                                       HAMN 103
      X(3)=LT(LATF+2)                                       HAMN 104
      IF (MD.EQ.0) GO TO 165                                HAMN 105
C     TABULAR DAY + NON TABULAR LATITUDE                    HAMN 106
      Y(1)=L((LATF-1)*16+MD)                                HAMN 107
      Y(2)=L(LATF*16+MD)                                    HAMN 108
      Y(3)=L((LATF+1)*16+MD)                                HAMN 109
      GO TO 170.                                            HAMN 110
C     NON TABULAR DATE + NON TABULAR LATITUDE               HAMN 111
  165 M1=MI-1                                               HAMN 112
      Y(1)=YLAG(DAY,DATE(M1),L((LATF-1)*16+M1),4)           HAMN 113
      Y(2)=YLAG(DAY,DATE(M1),L(LATF*16+M1),4)               HAMN 114
      Y(3)=YLAG(DAY,DATE(M1),L((LATF+1)*16+M1),4)           HAMN 115
  170 HAMN=YLAG(LAT,X,Y,3)                                  HAMN 116
      DAY=DAY+1.0                                           HAMN 117
  175 DAY=DAY+1.0                                           HAMN 118
  180 SUM=SUM+HAMN                                          HAMN 119
      HAMN=AMIN1(2729.0,AMAX1(SUM/DAYS,0.0))                HAMN 120
      IF (LEAP.NE.0) RETURN                                 HAMN 121
      DO 185 I=4,16                                         HAMN 122
      DATE(I)=DATE(I)-1.0                                   HAMN 123
  185 CONTINUE                                              HAMN 124
      DO 190 I=2,11                                         HAMN 125
      IM(I)=IM(I)-1                                         HAMN 126
      N(I)=N(I)-1                                           HAMN 127
  190 CONTINUE                                              HAMN 128
      RETURN                                                HAMN 129
      END                                                   HAMN 130
      FUNCTION YLAG(XI,X,Y,N)                               YLAG   2
C     N-POINT LAGRANGIAN INTERPOLATION WHERE I=1,N          YLAG   3
C     SPECIAL VERSION FOR USE WITH FUNCTION #HAMN#          YLAG   4
C     PROGRAM AUTHOR--E.C.LONG.  COMPUTER SCIENCES DIVISION--ORNL  YLAG   5
C     UNION CARBIDE NUCLEAR DIVISION.  OAK RIDGE, TENNESSEE YLAG   6
      DIMENSION X(N),Y(N)                                   YLAG   7
      S=0.0                                                 YLAG   8
      P=1.0                                                 YLAG   9
      DO 110 J=1,N                                          YLAG  10
      P=P*(XI-X(J))                                         YLAG  11
      D=1.0                                                 YLAG  12
      DO 105 I=1,N                                          YLAG  13
      IF (I.NE.J) GO TO 100                                 YLAG  14
      XD=XI                                                 YLAG  15
      GO TO 105                                             YLAG  16
  100 XD=X(J)                                               YLAG  17
  105 D=D*(XD-X(I))                                         YLAG  18
  110 S=S+Y(J)/D                                            YLAG  19
      YLAG=S*P                                              YLAG  20
      RETURN                                                YLAG  21
      END                                                   YLAG  22
```

Figure B.1 (Continued).

```
      PROGRAM COMET(INPUT,OUTPUT,TAPE5=INPUT,TAPE6=OUTPUT)
C
C     THIS PROGRAM CALCULATES THE DIFFERENCE BETWEEN THE EQUILIBRIUM TEMP-
C     ERATURES OF TWO DATA SETS AND THE SENSITIVITY TO THE VARIOUS PARA-
C     METERS.
C     R. CODELL AND W. NUTTLE, USNRC, OCTOBER, 1978
C
C     TD1= DEW POINT TEMP. FOR DATA SET 1    (F)
C     TA1= DRY BULB TEMP. FOR DATA SET 1     (F)
C     W1= WIND SPEED FOR DATA SET 1     (MPH)
C     H1= RATE OF INSOLATION FOR DATA SET 1     (BTU/FT**2/DAY)
C     TD2= DEW POINT TEMP. FOR DATA SET 2     (F)
C     TA2= DRY BULB TEMP. FOR DATA SET 2     (F)
C     W2= WIND SPEED FOR DATA SET 2     (MPH)
C     H2= RATE OF INSOLATION FOR DATA SET 2     (BTU/FT**2/DAY)
C
      COMMON/EVAP/AK,B
      DATA QX,QY,QX2,QY2,QCROSS/5*0.0/
      DATA ERR/1.0E-30/
      DATA SX,SY,SX2,SY2,SCROSS/5*0./
      WRITE(6,100)
100   FORMAT(1H1,10X,'PROGRAM TO COMPARE EQUILIBRIUM TEMPERATURES FROM T
     1WO DATA SETS AND COMPUTE THE SENSITIVITY OF EACH VARIABLE',//)
      READ(5,499)I
499   FORMAT(I2)
      DO 2 J=1,I
      READ(5,500) TD1,TA1,W1,H1,TD2,TA2,W2,H2
500   FORMAT(8F10.1)
      IF(H2.EQ.0.) H2=H1
C
C     CALCULATE EQUILIBRIUM TEMPERATURES
C
      E1=E(TD1,TA1,W1,H1)
      EVAP1=30.*(AK-15.7)*B*(E1-TD1)/(62.4*(.26+B)*1000)
      E2=E(TD2,TA2,W2,H2)
      EVAP2=30.*(AK-15.7)*B*(E2-TD2)/(62.4*(.26+B)*1000)
      DE=E2-E1
      DEVAP=EVAP2-EVAP1
      WRITE(6,99)
      WRITE(6,101)TD1,TA1,W1,H1,E1,EVAP1
      WRITE(6,200) TD2,TA2,W2,H2,E2,EVAP2
99    FORMAT(T26,'DEW POINT',T42,'DRY BULB',T56,'WIND SPEED',T69,
     1'SOLAR RAD.',T82,'EQUILIBRIUM TEMP.',T104,'EVAPORATION',//,T27,
     2'(DEG. F)',T59,'(MPH)',T67,'(BTU/FT**2/DY)',T86,'(DEG. F)',T102,
     3'(FT**3/FT**2)',//)
101   FORMAT(10X,'DATA SET 1',F12.2,4F15.2,F20.2,/)
200   FORMAT(10X,'DATA SET 2',F12.2,4F15.2,F20.2,//)
      WRITE(6,102) DE,DEVAP
102   FORMAT(T77,'E2-E1 = ', F6.3,5X,'EVAP2-EVAP1 = ',F7.2)
C
C     CALCULATE SUMS FOR CORRELATION COEFFICIENTS
C
      SX=SX+E1
      SX2=SX2+E1**2
      SY=SY+E2
      SY2=SY2+E2**2
      SCROSS=SCROSS+E1*E2
      QX=QX+EVAP1
      QX2=QX2+EVAP1**2
      QY=QY+EVAP2
      QY2=QY2+EVAP2**2
      QCROSS=QCROSS+EVAP1*EVAP2
```

Figure B.2 Listing of Program COMET.

```
C
C        DIFFERENCES IN EQUILIBRIUM TEMP DUE TO EACH PARAMETER.
C
         DTD=E(TD2,TA1,W1,H1)-E1
         DTA=E(TD1,TA2,W1,H1)-E1
         DW=E(TD1,TA1,W2,H1)-E1
         DH=E(TD1,TA1,W1,H2)-E1
         DTOT=DTD+DTA+DW+DH
         WRITE(6,5)
   5 FORMAT(//10X, 'DIFFERENCES IN E BETWEEN DATA SET 2 AND DATA SET 1
     1BY PARAMETER',/)
         WRITE(6,6)DTD
   6 FORMAT(10X,'DIFFERENCE DUE TO DEW POINT = ',T50,F10.3,' DEG. F')
         WRITE(6,7)DTA
   7 FORMAT(10X,'DIFFERENCE DUE TO DRY BULB TEMP. = ',T50,F10.3,' DEG.
     1F')
         WRITE(6,8) DW
   8 FORMAT(10X,'DIFFERENCE DUE TO WIND SPEED = ',T50,F10.3,' DEG. F')
         WRITE(6,9)DH
   9 FORMAT(10X,'DIFFERENCE DUE TO INSOLATION = ',T50,F10.3,' DEG. F')
         WRITE(6,10)DTOT
  10 FORMAT(10X,'SUMMATION OF INDIVIDUAL DIFFERENCES = ',T50,F10.3,' DE
     1G. F',//,1X,130('*'),///)
   2 CONTINUE
C
C        CORRELATION ANALYSIS
C
         SXX=I*SX2-SX**2
         SYY=I*SY2-SY**2
         SXY=I*SCROSS-SX*SY
         RSQ=(SXY**2+ERR)/(SXX*SYY+ERR)
         QXX=I*QX2-QX**2
         QYY=I*QY2-QY**2
         QXY=I*QCROSS-QX*QY
         QRSQ=(QXY**2+ERR)/(QXX*QYY+ERR)
         SERR=SQRT(((SXX*SYY)-SXY**2)/(I*(I-2)*SXX))
         QSERR=SQRT(((QXX*QYY)-QXY**2)/(I*(I-2)*QXX))
         WRITE(6,300) RSQ,SERR
         WRITE(6,310) QRSQ,QSERR
 300 FORMAT(10X,'SAMPLE R SQUARED FOR EQUILIBRIUM TEMP. = ',F10.3,
     1 10X,'STANDARD ERROR = ',F10.3,' DEG.F')
 310 FORMAT(10X,'SAMPLE R SQUARED FOR EVAPORATION = ', E13.3,
     1 10X,'STANDARD ERROR = ',E13.3,'FT**3/FT**2')
         SXXI=SX /I
         SYYI=SY /I
         BIAS=SYYI-SXXI
         WRITE(6,250) SXXI,SYYI,BIAS
 250 FORMAT(10X,'AVERAGE E, DATA SET 1 = ',F12.3,/,10X,'AVERAGE E, DATA
     1 SET 2 = ',F12.3,/,10X,'AVERAGE E2 - AVERAGE E1 = ',F12.4)
         EBIAS=(QY-QX)/I
         WRITE(6,251) EBIAS
 251 FORMAT(10X, 'AVERAGE EVAP2 - AVERAGE EVAP1 = ',F12.4)
         STOP
         END
         FUNCTION E(TD,TA,W,H)
C
C        CALCULATES THE EQUILIBRIUM TEMPERATURE BY THE BRADY METHOD IN
C        AN ITERATIVE PROCESS.
C
         COMMON/EVAP/AK,B
         DATA AK,ES/100.,100./
         DO 1 I=1,50
         TSTAR=(ES+TD)/2.



         B=.255-.0085*TSTAR+.000204*TSTAR**2
         AK=15.7+(B+.26)*(70.+.7*W**2)
         E=H/AK+(B*TD+.26*TA)/(B+.26)
         IF(ABS(ES-E).LT..001) GO TO 2
         ES=E
   1 CONTINUE
   2 RETURN
         END
```

Figure B.2 (Continued).

```
          PROGRAM UHS3 (INPUT,OUTPUT,TAPE5=INPUT,TAPE6=OUTPUT,
        1 TAPE8)                                                          000110
C         PROGRAM TO CALCULATE MAX TEMPERATURE IN A UHS POND              000120
C         BY PLUG,MIXED,AND STRATIFIED MODELS                            000130
C         R CODELL USNRC NOV 1978                                        000140
          DIMENSION R(10),S(10),B(10),C(11) , TIME(20),ITITLE(80)
          LOGICAL FLAG1                                                  000160
          COMMON AK1,E,E2,BETA,TSKIP,QBASE,FBASE,M1,M2,BTA,BTD,BHS,BW,
        1 IMET,BLOW,F1,Q1,TD,TA,HS,W,G(1000,4),HEAT(20),FLOW(20),TH(20),
        2 NMET,NH,A,DTMET
          NAMELIST/HFT/ NH,HEAT,FLOW,TH                                  000235
          DATA M4,NSTEPS,NPRINT/0,100,10/
          DATA DT,TZERO/0.2,80.0/
          DATA TIMEM,TIMEST,TIMEPL/3*0.0/
          NAMELIST /INLIST/ VZERO,BLOW,A,NH,NSTEPS,NPRINT,DT,TZERO,DTMET  000240
        1 ,TSKIP,QBASE,FBASE,E,AK1,IMET,AMAKE
        1 ,BTA,BTD,BHS,BW                                                000255
        2 ,HEAT,FLOW,TH                                                  000256
          READ(5,101) NMET
  101 FORMAT(I5)                                                         000280
C         READ IN MET TABLE(WIND SP.,DRY BULB,DEW PT,SOL RAD)            000285
          READ(5,1) (G(I,4),G(I,2),G(I,1),G(I,3),CC,RH,I=1,NMET)
        1 FORMAT(3X,3F5.0,F6.0,2F4.0,3F5.0,F6.0,2F4.0)
C         VZERO = VOLUME OF POND FT**3
C         BLOW = BLOWDOWN RATE OUT FT**3/HR
C         A = SURFACE AREA FT**2
C         NSTEPS = NUMBER OF INTEGRATION STEPS
C         NPRINT = PRINT EVERY NPRINT STEPS
C         DT = INTEGRATION TIMESTEP, HRS
C         TZERO = INITIAL POND TEMP DEG.F
C         G(I,1)=TD=DEW POINT, DEG.F
C         G(I,2)=TA=DRY BULB DEG.F
C         G(I,3) =HS =  SOLAR RADIATION BTU/(FT**2 DAY)
C         G(I,4)= W = WIND SPEED MPH
C         TSKIP = DELAY START OF HEAT TABLE BY TSKIP HRS
C         QBASE = BASE HEAT LOAD, BTU/HR
C         FBASE = BASE FLOW, FT**3/HR
C         E CONST EQUILIBRIUM TEMP, DEG.F IF USED
C         AK1 = CONSTANT H.T.COEFF, BTU/(FT**2 DAY DEG.F), IF USED
C         IMET = OPTIONAL CONSTANT E AND AK1 IF IMET = 1
C         BTA,BTD,BHS,BW = BIASES TO BE ADDED TO ALL MET TABLE VALUES
C                          OF TA,TD,HS,AND W RESPECTIVELY
C         NH = NUMBER OF ENTRIES IN HEAT TABLE
C         HEAT = ARRAY OF HEAT INPUTS, BTU/HR
C         FLOW = ARRAY OF FLOW RATES, FT**3/HR
C         TH = ARRAY OF CORRESPONDING TIMES FOR HEAT AND FLOW ARRAYS
          BLOW=0                                                         000310
          AMAKE=0
          DTMET=1                                                        000360
          TSKIP=0                                                        000370
          QBASE=0                                                        000380
          FBASE=0                                                        000390
          BTA=0                                                          000393
          BTD=0                                                          000394
          BHS=0                                                          000395
          BW=0                                                           000396
          E=80                                                           000400
          AK1=150                                                        000410
          IMET=0                                                         000420
          TD=G(1,1)                                                      000430
          TA=G(1,2)                                                      000440
          HS=G(1,3)                                                      000450
          W=G(1,4)                                                       000460
          READ(5,HFT)
          DO 4 I=1,NH                                                    000520
          TIME(I)=TH(I)
          TH(I)=TH(I)+1.0E-20                                            000540
        4 TH(I)=ALOG(TH(I))                                              000550
          IF(NH.GT.1) GOTO 710                                          000560
          FLOW(2)=FLOW(1)                                                000570
          HEAT(2)=HEAT(1)                                                000580
          NH=2                                                          000590
          TH(2)=1.0E6                                                    000600
  710 CONTINUE                                                          000610
```

Figure B.3 Listing of Program UHS3.

101

```
6000 CONTINUE                                                        000620
     READ(5,480)ITITLE
 480 FORMAT(80A1)
     READ(5,INLIST)
     IF(VZERO.LE.0.0) STOP                                           000640
     WRITE(6,490) ITITLE
 490 FORMAT(1H1,5(/),T20,80A1)
     WRITE(6,500) VZERO,A,BLOW,AMAXE,NSTEPS,NPRINT,DT,TZERO,
    1TSKIP,QBASE,FBASE,E,AK1,IMET,BTA,BTD,BHS,BW
 500 FORMAT(    5(/),T43,'VZERO',T57,'A',T66,'BLOW',T76,'AMAXE',/, T38,
    1E11.5,1X,E11.5,3X,E9.3,1X,E9.3,//,T43,'NSTEPS',T53,'NPRINT',T65,
    2'DT',T73,'TZERO',T84,'TSKIP',/,T43,I5,T54,I4,T64,F5.3,T73,F5.1,
    3T83,F6.1,//,T44,'QBASE',T54,'FBASE',T65,'E',T74,'AK1',T84,'IMET',/
    4,T41,E9.3,1X,E9.3,3X,F5.1,5X,F5.1,7X,I1,//,T45,'BTA',T55,'BTD',T64
    5,'BHS',T75,'BW',//,T44,F4.1,6X,F4.1,5X,F6.1,5X,F4.1,6(/),T43,
    635('.'),//,T43,': HEAT IN : TIME FROM : FLOW IN :',/,T43,':  BTU/
    7HR :   START   : FT**3/HR :',/,T43,35('.'))
     DO 2 I=1,NH
   2 WRITE(6,510)HEAT(I),TIME(I),FLOW(I)
 510 FORMAT(T43,':',   E9.3,1X,':',2X,F7.2,2X,':',   E9.3,1X,':')
     WRITE(6,520)
 520 FORMAT(T43,35('.'),5(/),T41,13('*'),' MODEL RESULTS ',13('*'),///,
    1T38,'..TIME..........TEMPERATURE (F).........VOLUME....',/,T38,':  HR
    2  : MIXED : STRAT : PLUG  :    FT**3  :',/,T38,46('.'))
     FLAG1=.FALSE.                                                   000645
     TS=0                                                            000650
     TMAXST=0                                                        000660
     TMAXPL=0                                                        000670
     M1=1                                                            000680
     M2=1                                                            000690
     X=.001                                                          000700
     DO 3 I=1,10                                                     000710
     S(I)=TZERO                                                      000720
   3 C(I+1)=TZERO                                                    000730
     T=TZERO                                                         000740
     V=VZERO                                                         000745
C    BEGIN NUMERICAL INTEGRATIONS                                    000780
     DO 6 M=1,NSTEPS                                                 000790
C    MIXED TANK SOLUTIONS                                            000800
     CALL MIXED(F2,F3,T,V,X)                                         000810
     CALL MIXED(F7,F8,T+DT*F2,V+DT*F3,X+DT)                          000820
     T=T+DT*(F2+F7)/2                                                000830
     V=V+DT*(F3+F8)/2                                                000840
C    FIND MAX TEMPERATURE FOR MIXED MODEL                            000850
     IF(T.LT.TS) GOTO 63                                            000860
     TS=T                                                            000870
     TIMEM=X                                                         000880
  63 CONTINUE                                                        000890
     M4=M4+1                                                         000900
C    STRATIFIED MODEL                                                000910
     AL1=V/A                                                         000920
     AL3=V/10                                                        000930
     AL4=AL1/10                                                      000940
     AL6=DT/(62.4*AL3)                                               000950
     AL2=F1/A                                                        000960
     AL5=AL2*DT/AL4                                                  000970
     CALL EQTEMP(S(1))                                               000980
     AK=AK1*A/24                                                     000990
     R(1)=S(1)+AL5*(S(10)-S(1))+(Q1-AK*(S(1)-E))*AL6                 001000
     DO 9 I=2,10                                                     001010
   9 R(I)=S(I)+AL5*(S(I-1)-S(I))                                     001020
     DO 10 I=1,10                                                    001030
  10 S(I)=R(I)                                                       001040
C    PLUG FLOW MODEL                                                 001140
     C(1)=C(11)                                                      001150
     DO 20 I=1,10                                                    001160
     B(I)=C(I+1)+AL5*(C(I)-C(I+1))                                   001170
     CALL EQTEMP(C(I))                                               001180
     AK=AK1*A/240                                                    001190
  20 B(I)=B(I)-AK*(B(I)-E)*AL6                                       001200
     B(1)=B(1)+AL6*Q1                                                001210
     DO 21 I=1,10                                                    001220
  21 C(I+1)=B(I)                                                     001230
     IF(S(10).LT.TMAXST) GOTO 61                                     001240
     TMAXST=S(10)                                                    001250
     TIMEST=X                                                        001260
```

Figure B.3 (Continued).

```
      61 CONTINUE                                                     001270
         IF(C(11).LT.TMAXPL) GOTO 62                                  001280
         TMAXPL=C(11)                                                 001290
         TIMEPL=X                                                     001300
      62 CONTINUE                                                     001310
         X=X+DT                                                       001320
         IF(NPRINT.GT.M4) GOTO 6                                      001330
         M4=0                                                         001340
         WRITE(6,51) X,T,S(10),C(10),V
      51 FORMAT(T38,'=',4(1X,F5.1,1X,'='),E11.5,1X,'=')
       6 CONTINUE                                                     001370
         WRITE(6,55) TS,TIMEM,TMAXST,TIMEST,TMAXPL,TIMEPL
      55 FORMAT(T38,46('.'),///,T40,'MAXIMUM MODELLED TEMPERATURES:',/,T40
        1,'MIXED MODEL = ',F8.2,' AT ',F8.2 ' HOURS',/,T40,'STRAT MODEL = '
        2,F8.2,' AT ',F8.2,' HOURS',/,T40, 'PLUG  MODEL = ',F8.2,' AT ',
        3F8.2,' HOURS')
         GOTO 6000                                                    001430
         END                                                          001440
         SUBROUTINE MIXED(FA,FB,T,V,X)                                001450
C     MIXED TANK MODEL
         COMMON AK1,E,E2,BETA,TSKIP,QBASE,FBASE,M1,M2,BTA,BTD,BHS,BW,
        1 IMET,BLOW,F1,Q1,TD,TA,HS,W,G(1000,4),HEAT(20),FLOW(20),TH(20),
        2 NMET,NH,A,DTMET
C     LOG-LINEAR INTERPOLATION OF HEAT TABLE                          001530
         DO 1 M1=M2,NH                                                001540
         X1=X-TSKIP                                                   001550
         IF(X1.LE.0.0) X1=.00001                                      001560
         X9=ALOG(X1)                                                  001570
         IF(X9.LT.TH(M1)) GOTO 1                                      001580
         IF(X9.LT.TH(M1+1)) GOTO 1210                                 001590
       1 CONTINUE                                                     001600
    1210 F4=(X9-TH(M1))/(TH(M1+1)-TH(M1))                             001610
         M2=M1                                                        001620
C     EXTERNAL HEAT INPUT TO POND
         Q1=HEAT(M1)+F4*(HEAT(M1+1)-HEAT(M1))                         001630
C     CIRCULATION THROUGH POND
         F1=FLOW(M1)+F4*(FLOW(M1+1)-FLOW(M1))                         001640
C     ADD BASE HEAT LOAD AND FLOW, IF ANY
         Q1=Q1+QBASE                                                  001650
         F1=F1+FBASE                                                  001660
C     LINEAR INTERPOLATION OF MET TABLE                               001670
         IF(NMET.EQ.1) GOTO 100                                       001680
         M1=X/DTMET+1                                                 001690
         F4=(X-(M1-1)*DTMET)/DTMET                                    001700
         TD=G(M1,1)+F4*(G(M1+1,1)-G(M1,1))                            001710
         TA=G(M1,2)+F4*(G(M1+1,2)-G(M1,2))                            001720
         HS=G(M1,3)+F4*(G(M1+1,3)-G(M1,3))                            001730
         W=G(M1,4)+F4*(G(M1+1,4)-G(M1,4))                             001740
         TD=TD+BTD                                                    001742
         TA=TA+BTA                                                    001743
         HS=HS+BHS                                                    001744
         W=W+BW                                                       001745
     100 CONTINUE                                                     001750
         CALL EQTEMP(T)                                               001760
         AK=AK1*A/24                                                  001770
C     RATE OF TEMPERATURE CHANGE, DEG F/HR
         FA=(Q1-AK*(T-E))/(62.4*V)                                    001780
C     EVAPORATION RATE, FT**3/HR
         E2=(AK1-15.7)*BETA*(T-TD)*A/(62.4*(.26+BETA)*24000)          001790
C     RATE OF VOLUME CHANGE, FT**3/HR
         FB=-BLOW-E2                                                  001810
         RETURN                                                       001820
         END                                                          001830
         SUBROUTINE EQTEMP(T)                                         001840
C     CALCULATE EQUILIBRIUM TEMPERATURE AND HEAT TRANSFER COEFF
         COMMON AK1,E,E2,BETA,TSKIP,QBASE,FBASE,M1,M2,BTA,BTD,BHS,BW,
        1 IMET,BLOW,F1,Q1,TD,TA,HS,W,G(1000,4),HEAT(20),FLOW(20),TH(20),
        2 NMET,NH,A,DTMET
         IF(IMET.EQ.1) RETURN                                         001920
C     WIND FUNCTION
         G7=70+.7*W**2                                                001930
         G5=(TD+T)/2                                                  001940
         BETA=.255-.0085*G5+.000204*G5**2                            001950
C     SURFACE HEAT TRANSFER                                           001960
         AK1=15.7+(.26+BETA)*G7                                       001970
         E=HS/AK1+(.26*TA+BETA*TD)/(.26+BETA)                        001980
         RETURN                                                       001990
         END                                                          002000
```

Figure B.3 (Continued).

103

```
      SUBROUTINE PSY1(DB,WB,PB,DP,PV,W,H,V,RH)                    PSY1    1
C     THIS ROUTINE CALCULATES' VAPOR PRESSURE PV, HUMIDITY RATIO W,  PSY1    2
C         ENTHALPY H, VOLUME V, RELATIVE HUMIDITY RH, AND            PSY1    3
C         DEW POINT TEMPERATURE DP\                                  PSY1    4
C         WHEN THE DRY BULB TEMPERATURE DB, WET BULB TEMPERATURE WB, PSY1    5
C         AND BAROMETRIC PRESSURE PB ARE GIVEN                       PSY1    6
C     UNITS' DB, WB, + DP )F>\ PB, + PV )IN OF HG>\ W)= WATER VAPOR   PSY1    7
C         PER = DRY AIR>\ H )BTU/= OF DRY AIR>\ V )FT**3/= OF DRY     PSY1    8
C         AIR\ RH IS A FRACTION, NOT (                               PSY1    9
      C(F)=(F-32.0E0)/1.8E0                                         PSY1   10
      PVP=PVSF(WB)                                                  PSY1   11
      WSTAR=0.622*PVP/(PB-PVP)                                      PSY1   13
      IF (WB.GT.32.0) GO TO 105                                     PSY1   14
      PV=PVP-5.704E-4*PB*(DB-WB)/1.8                                PSY1   15
      GO TO 110                                                     PSY1   16
  100 PV=PVP                                                        PSY1   17
      GO TO 110                                                     PSY1   18
  105 CDB=C(DB)                                                     PSY1   19
      CWB=C(WB)                                                     PSY1   20
      HL=597.31+0.4409*CDB-CWB                                      PSY1   21
      CH=0.2402+0.4409*WSTAR                                        PSY1   22
      EX=(WSTAR-CH*(CDB-CWB)/HL)/0.622                              PSY1   23
      PV=PB*EX/(1.+EX)                                              PSY1   24
  110 W=0.622*PV/(PB-PV)                                            PSY1   25
      V=0.754*(DB+459.7)*(1.0+7000.0*W/4360.0)/PB                   PSY1   26
      H=0.24*DB+(1061.0+0.444*DB)*W                                 PSY1   27
      IF (PV.GT.0.0) GO TO 115                                      PSY1   28
      PV=0.0                                                        PSY1   29
      DP=0.0                                                        PSY1   30
      RH=0.0                                                        PSY1   31
      RETURN                                                        PSY1   32
  115 IF (DB.NE.WB) GO TO 120                                       PSY1   33
      DP=DB                                                         PSY1   34
      RH=1.0                                                        PSY1   35
      RETURN                                                        PSY1   36
  120 DP=DPF(PV)                                                    PSY1   37
      RH=PV/PVSF(DB)                                                PSY1   38
      RETURN                                                        PSY1   39
      END                                                           PSY1   40
      SUBROUTINE PSY2(DB,DP,PB,WB,PV,W,H,V,RH)                      PSY2    1
C     THIS ROUTINE CALCULATES' WET BULB TEMPERATURE WB, HUMIDITY    PSY2    2
C         RATIO W, ENTHALPY H, VOLUME V, VAPOR PRESSURE PV,         PSY2    3
C         AND RELATIVE HUMIDITY RH\                                 PSY2    4
C         WHEN DRY BULB TEMPERATURE DB, DEW POINT TEMPERATURE DP,   PSY2    5
C         AND BAROMETRIC PRESSURE PB ARE GIVEN                      PSY2    6
C     UNITS' DB, WB, + DP )F>\ PB, + PV )IN OF HG>\ W)= WATER VAPOR  PSY2    7
C         PER = DRY AIR>\ H )BTU/= OF DRY AIR>\ V )FT**3/= OF DRY    PSY2    8
C         AIR\ RH IS A FRACTION, NOT (                              PSY2    9
      IF (DP.GT.DB) DP=DB                                           PSY2   10
      PV=PVSF(DP)                                                   PSY2   11
      PVS=PVSF(DB)                                                  PSY2   12
      RH=PV/PVS                                                     PSY2   13
      W=0.622*PV/(PB-PV)                                            PSY2   14
      V=0.754*(DB+459.7)*(1.0+7000.0*W/4360.0)/PB                   PSY2   15
      H=0.24*DB+(1061.0+0.444*DB)*W                                 PSY2   16
      IF (H.GT.0.0) GO TO 100                                       PSY2   17
      WB=DP                                                         PSY2   18
      RETURN                                                        PSY2   19
  100 WB=WBF(H,PB)                                                  PSY2   20
      RETURN                                                        PSY2   21
      END                                                           PSY2   22
      FUNCTION PVSF(X)                                              PVSF    1
      DIMENSION A(6),B(4),P(4)                                      PVSF    2
      DATA A/-7.90298,5.02808,-1.3816E-7,11.344,8.1328E-3,-3.49149/ PVSF    3
      DATA B/-9.09718,-3.56654,0.876793,0.0060273/                  PVSF    4
      T=(X+459.688)/1.8                                             PVSF    5
      IF (T.LT.273.16) GO TO 100                                    PVSF    6
      Z=373.16/T                                                    PVSF    7
      P(1)=A(1)*(Z-1.0)                                             PVSF    8
      P(2)=A(2)*ALOG10(Z)                                           PVSF    9
      Z1=A(4)*(1.0-1.0/Z)                                           PVSF   10
      P(3)=A(3)*(10.0**Z1-1.0)                                      PVSF   11
```

Figure B.4 Listing of Psychrometric Subroutines.

104

```
            Z1=A(6)*(Z-1.0)                                    PVSF  12
            P(4)=A(5)*(10.0**Z1-1.0)                           PVSF  13
            GO TO 105                                          PVSF  14
        100 Z=273.16/T                                         PVSF  15
            P(1)=B(1)*(Z-1.0)                                  PVSF  16
            P(2)=B(2)*ALOG10(Z)                                PVSF  17
            P(3)=B(3)*(1.0-1.0/Z)                              PVSF  18
            P(4)=ALOG10(B(4))                                  PVSF  19
        105 SUM=0.0                                            PVSF  20
            DO 110 I=1,4                                       PVSF  21
        110 SUM=SUM+P(I)                                       PVSF  22
            PVSF=29.921*10.0**SUM                              PVSF  23
            RETURN                                             PVSF  24
            END                                                PVSF  25
            FUNCTION DPF(PV)                                   DPF    1
    C       THIS ROUTINE CALCULATES DEW-POINT TEMPERATURE FOR A GIVEN  DPF    2
    C           VAPOR PRESSURE PV                              DPF    3
            DP(A,B,C,Y)=A+(B+C*Y)*Y                            DPF    4
            Y=ALOG(PV)                                         DPF    5
            IF (PV.GT.0.1836) GO TO 100                        DPF    6
            DPF=DP(71.98,24.873,0.8927,Y)                      DPF    7
            RETURN                                             DPF    8
        100 DPF=DP(79.047,30.579,1.8893,Y)                     DPF    9
            RETURN                                             DPF   10
            END                                                DPF   11
            FUNCTION WBF(H,PB)                                 WBF    1
    C       THIS ROUTINE APPROXIMATES THE WET BULB TEMPERATURE FROM    WBF    2
    C           ENTHALPY H, AND BAROMETRIC PRESSURE PB         WBF    3
            WB(A,B,C,D,Y)=A+(B+(C+D*Y)*Y)*Y                    WBF    4
            W(PV,PB)=0.622*PV/(PB-PV)                          WBF    5
            X(WB12,W12)=0.24*WB12+(1061.0+0.444*WB12)*W12      WBF    6
            IF (H.LE.0.0) GO TO 105                            WBF    7
            Y=ALOG(H)                                          WBF    8
            IF (H.GT.11.758) GO TO 100                         WBF    9
            WBF=WB(0.6041,3.4841,1.3601,0.97307,Y)             WBF   10
            RETURN                                             WBF   11
        100 WBF=WB(30.9185,-39.682,20.5841,-1.758,Y)           WBF   12
            RETURN                                             WBF   13
        105 WB1=150.0                                          WBF   14
            PV1=PVSF(WB1)                                      WBF   15
            W1=W(PV1,PB)                                       WBF   16
            X1=X(WB1,W1)                                       WBF   17
            Y1=H-X1                                            WBF   18
        110 WB2=WB1-1.0                                        WBF   19
            PV2=PVSF(WB2)                                      WBF   20
            W2=W(PV2,PB)                                       WBF   21
            X2=X(WB2,W2)                                       WBF   22
            Y2=H-X2                                            WBF   23
            IF (Y1*Y2) 130,120,115                             WBF   24
        115 WB1=WB2                                            WBF   25
            Y1=Y2                                              WBF   26
            GO TO 110                                          WBF   27
        120 IF (Y1.NE.0.0) GO TO 125                           WBF   28
            WBF=WB1                                            WBF   29
            RETURN                                             WBF   30
        125 WBF=WB2                                            WBF   31
            RETURN                                             WBF   32
        130 Z=ABS(Y1/Y2)                                       WBF   33
            WBF=(WB2*Z+WB1)/(1.0+Z)                            WBF   34
            RETURN                                             WBF   35
            END                                                WBF   36
```

Figure B.4 (Continued).

*U.S. GOVERNMENT PRINTING OFFICE: 1980-0-341-742/581

NRC FORM 335 (7 77)	U.S. NUCLEAR REGULATORY COMMISSION BIBLIOGRAPHIC DATA SHEET		1. REPORT NUMBER (Assigned by DDC) NUREG-0693
4. TITLE AND SUBTITLE (Add Volume No., if appropriate) Analysis of Ultimate Heat Sink Cooling Ponds			2. (Leave blank)
			3. RECIPIENT'S ACCESSION NO.
7. AUTHOR(S) Richard B. Codell and William K. Nuttle			5. DATE REPORT COMPLETED MONTH July | YEAR 1980
9. PERFORMING ORGANIZATION NAME AND MAILING ADDRESS (Include Zip Code) Office of Nuclear Reactor Regulation U.S. Nuclear Regulatory Commission Washington, DC 20555			DATE REPORT ISSUED MONTH November | YEAR 1980
			6. (Leave blank)
			8. (Leave blank)
12. SPONSORING ORGANIZATION NAME AND MAILING ADDRESS (Include Zip Code) Same as 9, above.			10. PROJECT/TASK/WORK UNIT NO. N/A
			11. CONTRACT NO. N/A
13. TYPE OF REPORT Technical Report		PERIOD COVERED (Inclusive dates) N/A	
15. SUPPLEMENTARY NOTES Computer Programs Available			14. (Leave blank)

16. ABSTRACT (200 words or less)

A method to analyze the performance of ultimate heat sink cooling ponds is presented. A simple mathematical model of a cooling pond is used to scan weather data to determine the period of the record for which the most adverse pond temperature or rate of evaporation would occur. Once the most adverse conditions have been determined, the peak pond temperature can be calculated. Several simple mathematical models of ponds are described; these could be used to determine peak pond temperature, using the identified meteorological record. Evaporative water loss may be found directly from the scanning by a simple and conservative heat-and-material balance.

Methodology by which short periods of onsite data can be compared with longer offsite records is developed, so that the adequacy of the offsite data for pond performance computations can be established.

17. KEY WORDS AND DOCUMENT ANALYSIS   17a. DESCRIPTORS

Ultimate Heat Sink
Cooling Ponds
Mathematical Models
Evaporation

17b. IDENTIFIERS/OPEN ENDED TERMS

18. AVAILABILITY STATEMENT Unclassified	19. SECURITY CLASS (This report) Unclassified	21. NO. OF PAGES
	20. SECURITY CLASS (This page) Unclassified	22. PRICE S

NRC FORM 335 (7 77)